Electric Power Planning for Regulated and Deregulated Markets

The Wiley Bicentennial–Knowledge for Generations

Each generation has its unique needs and aspirations. When Charles Wiley first opened his small printing shop in lower Manhattan in 1807, it was a generation of boundless potential searching for an identity. And we were there, helping to define a new American literary tradition. Over half a century later, in the midst of the Second Industrial Revolution, it was a generation focused on building the future. Once again, we were there, supplying the critical scientific, technical, and engineering knowledge that helped frame the world. Throughout the 20th Century, and into the new millennium, nations began to reach out beyond their own borders and a new international community was born. Wiley was there, expanding its operations around the world to enable a global exchange of ideas, opinions, and know-how.

For 200 years, Wiley has been an integral part of each generation's journey, enabling the flow of information and understanding necessary to meet their needs and fulfill their aspirations. Today, bold new technologies are changing the way we live and learn. Wiley will be there, providing you the must-have knowledge you need to imagine new worlds, new possibilities, and new opportunities.

Generations come and go, but you can always count on Wiley to provide you the knowledge you need, when and where you need it!

William J. Pesce
President and Chief Executive Officer

Peter Booth Wiley
Chairman of the Board

Electric Power Planning for Regulated and Deregulated Markets

Arthur Mazer

IEEE

IEEE PRESS

BICENTENNIAL 1807 WILEY 2007 BICENTENNIAL

WILEY-INTERSCIENCE
A John Wiley & Sons, Inc., Publication

Published by John Wiley & Sons, Inc., Hoboken, New Jersey
Published simultaneously in Canada

Wiley Bicentennial Logo: Richard J. Pacifico

Printed in the United States of America

Library of Congress Cataloging-in-Publication Data
Mazer, Arthur, 1958–
Electric Power Planning for Regulated and Deregulated Markets / by Arthur Mazer.
p. cm.
ISBN: 978-0-470-11882-5
1. Electricutilities—Management. 2. Risk management. 3. Competition. I. Title.
HD9685.A2M38 2007
333.793′20681—dc22

2006033932

10 9 8 7 6 5 4 3 2 1

To Mom, Linda, Lijuan, Julius and Amelia

Contents

Preface

Why write this text and give up so much playtime with my children? One of the pleasures of my career in planning and risk management roles with utilities and energy marketers has been the hiring of individuals new to the industry and watching them embark in new career directions. For some individuals this has been their first job. Others have transitioned into a new career field. Most are energetic, excited, and eager to learn. Every new hire has asked for introductory literature on the field. I have done a bit of searching and have been unable to find something that hits the mark. Risk management and planning teams are interdisciplinary. What is needed is a core set of material that is understood across several disciplines. What is available are excellent technical expositions that are only accessible to those with specialized education in a given discipline. My motivation is to fill the gap.

In filling this gap, the text provides an industry overview for many individuals who are in the private and public power service sectors as well as academicians wishing to investigate the industry. Companies and agencies operating in the power sector require combined skill sets of several disciplines including engineers, lawyers, systems specialists, economists, financial analysts, policy analysts, and applied mathematicians. The interaction of these professionals across their respective disciplines is key to the success of the company or agency. This text provides a fundamental pool of material that is required by these diverse professionals to interact effectively. The exposition is at a level that is accessible to a general audience.

Traditionally, utilities have assumed the role of planning and operating power systems on behalf of their customers. Utilities have developed expertise in these areas. As the industry moves away from a utility environment, other market participants are assuming the functions that utilities had. The Independent System Operator (ISO) operates the transmission network. Independent Power Producers (IPPs) build, own, and operate plants. Retailers procure load on behalf of a customer base. Within a traditionally regulated utility environment, these responsibilities are within the purview of utility planning. Within the new environment, they fall under new labels: asset management, investment analysis, and risk management. The main premise of this text is that new market participants must learn and understand traditional planning functions within a utility environment in order to properly perform their tasks.

A central objective of the text is the explanation of physical and financial equilibrium, balancing supply with demand, through decisions over different time frames. We provide a description of decision making in both regulated utility environments and market environments where there is customer choice. The text's organization follows from the premise that knowledge of the power delivery chain as characterized in a regulated utility environment is a prerequisite to understanding asset management, investment analysis, and risk management in a market environment. Accordingly, we present a complete discussion of utility operations and planning before addressing market environments.

Some of the material is technical and lends itself to mathematical analysis. The philosophy in presenting mathematical formalism is that it must be explanatory: Formulas must explain phenomena in a way that is superior to standard text or highlight information contained in text. Because we use mathematical formulations as an explanatory vehicle, it is possible to keep the material accessible to a broad audience. The material only assumes mathematical knowledge at the level of a standard high school curriculum and does not extend to calculus. We do not delve into solution techniques that would require a stronger mathematical background.

We are careful to ensure that the main body of the text is self contained. There are no prerequisites to the text. Nevertheless, there are occasional remarks that require a stronger background than the main body. It is left to the reader to determine his or her own level of engagement with the text. The reader may skip over the remarks without losing access to the remainder of the text or may delve into the details of every remark.

While there is a degree of mathematical formalism, the bulk of the material is not mathematical. There are descriptions of generation equipment, grid equipment, operations, operational objectives, market design, project management, risk management objectives, and other material that contain no mathematical formalism. The text also includes past newsworthy material insofar as the material enhances an understanding of operations and management. These newsworthy items include explanations of what went wrong at Three Mile Island, the Northeast blackout of 2003, and the California energy crisis.

Chapters 1 through 6 describe planning in a traditional utility environment, while Chapters 7 through 11 describe asset management, investment analysis, and risk management in a market environment. The specific topics of the chapters are as follows.

- Chapter 1 provides an overview of the subject as well as the text. This is a necessary starting point for those without an industry background as it provides context to the rest of the book.
- Chapter 2 describes physical characteristics of load and generation equipment. These physical characteristics influence the outcome of all planning and market activities.
- Chapter 3 describes transmission networks and operations and completes the description of physical factors influencing planning.
- Chapter 4 presents utility dispatch scheduling of generation units to meet the following day's load. While this chapter addresses short-term planning, it is

central to all planning activities and referred to repeatedly throughout the book.

- Chapter 5 describes a utility's process for determining infrastructure additions, both power and transmission, to the power delivery system.
- Chapter 6 describes a utility's management of existing generation and transmission assets through the setting of maintenance schedules, use of emissions-limited resources, and interutility contracting.
- Chapter 7 provides a description of a market environment. The remainder of the book discusses market activities within this environment.
- Chapter 8 describes the bidding and selection process that determines prices and the dispatch of units in a market environment.
- Chapter 9 describes the business environment and decision making for investing in power plants within a market environment.
- Chapter 10 describes financial risk management activities of various market participants.
- Chapter 11 contrasts the philosophies of two market structures, the California market that collapsed in 2001 and the more successful market known as PJM.

The utility activities of Chapters 4, 5, and 6 correspond to market-based activities in Chapters 8, 9, and 10.

Acknowledgments

It's a pleasure to have the opportunity to acknowledge many generous colleagues who have provided valuable input into the text.

I have the fortune of working with Milan Bjelogrlic. He has always made himself accessible for me to ask questions in areas where he is a leading authority. Lucky for me, Milan is a leading authority in every aspect of power generation and grid management. Jamal Jafari recommended standard texts for learning about grid management. He was always willing to review the material with me. Dennis Dayne is able to talk for hours about Hoover Dam. And I am able to listen for hours. Mike Borghi, Corinne Chandler, Sharim Chaudhury, Richard Goldberg, Beena Morar, Nathan Nguyen, and Inga Volkhonska reviewed the text and made suggestions for improvement.

If the reader finds the text insightful, it's because I listened to my colleagues. If there are areas that are off the mark, it's because I didn't listen well enough.

Thanks also go to the people at Wiley; in particular, George Telecki and Rachel Witmer provided much needed assistance and were very patient throughout a lengthy process.

Figure Citations

Two sources provided figures that are used in the text.

1. TVA. Figures were downloaded from their website, www.tva.com
2. The Final Report on the August 14, 2003 Blackout in the United States and Canada

The report was released by The Joint US-Canada Power System Outage Task Force. The following identifies the figure and source.

Figure	Source
2.5	TVA
2.6	TVA
2.7	TVA
2.8	TVA
2.9	TVA
2.10	TVA
2.11	TVA
3.8	Joint Task Force
3.9	Joint Task Force

About the Author

Art Mazer manages a group of analysts within the energy procurement department of Southern California Edison (SCE). The primary responsibility of the group is to develop methodologies and tools for valuation and selection of contracts in competitive solicitations that SCE conducts to acquire electric energy, capacity, and fuel.

Chapter 1

Overview

The business structure of the power industry is very simple in a traditionally regulated environment. Vertically integrated electric utilities oversee the entire chain of power delivery, while state commissions set customer rates. After the successful deregulation of airlines and telecommunications industries there has been a movement toward deregulating the electric power industry. The market model that is still evolving is considerably more complex than the traditionally regulated environment. There is a vision that the market should segment the responsibilities of the vertically integrated utilities along logical cross sections and different market participants would assume responsibility for a particular aspect of the power delivery chain. Central to this book is the premise that if one wishes to design and operate a market for the electric power industry that segments the delivery chain, one must first understand the planning and operations of vertically integrated utilities, with an emphasis on how the different elements of the delivery chain interact with one another.

This chapter begins with an overview of the roles and responsibilities of the utility within a traditionally regulated environment. Then the chapter looks at the market segmentation of the power delivery chain along with the different market participants that assume the various responsibilities. The chapter provides a first glimpse at topics that are discussed more fully in the remainder of the book and identifies chapters that provide further discussion of the topics. Respecting the premise noted at the end of the preceding paragraph, before addressing a market environment we first address the aspects of power delivery in a vertically integrated utility in both this chapter and the remainder of the book.

1.1 THE POWER DELIVERY CHAIN IN A VERTICALLY INTEGRATED UTILITY

This section describes the power delivery chain that a vertically integrated utility oversees. A brief description of the chain elements is provided, along with references

Electric Power Planning for Regulated and Deregulated Markets. By Arthur Mazer

to chapters in which a fuller discussion is present. There are three major elements in the chain associated with different planning horizons: long term, short term, and midterm.

1.1.1 Infrastructure Additions: Long-Term Planning

Electric power delivery requires an expensive infrastructure of generation and transmission equipment. The infrastructure must be expanded to meet growing demand for electric energy. In a regulated environment, utilities propose infrastructure development projects to state utility commissions. Project proposals should provide convincing evidence that the project economically addresses a community need. State commissions approve projects along with a customer rate structure that provides funding for the approved projects (funding includes a profit margin for the utility). Utilities then oversee the construction of new facilities.

An important aspect of infrastructure development is the interdependence of requirements between transmission-related equipment and generation equipment. The network of generation plants and transmission equipment is called the grid. Generation additions must be coordinated with other accompanying upgrades to the grid by an integrated approach. An integrated approach is necessary to address two objectives of infrastructure additions: cost minimization and reliability.

Chapter 2 characterizes load and load growth and additionally presents generation technologies that are available for growing the infrastructure. Chapter 3 addresses the configuration and operations of the grid that must be considered when expanding infrastructure. Chapter 5 describes the long-term planning and development process.

1.1.2 Day-Ahead Scheduling and Real-Time Grid Management: Short-Term Planning

Utilities must demonstrate effective use of generation units. This entails dispatching units that most economically serve customer load while also addressing security and reliability constraints imposed by the transmission system. Utilities set a day-ahead dispatch schedule against a forecasted load, verify that the transmission system can manage the forecasted power flows, and test the proposed dispatch against possible equipment failure to ensure security and reliability in the event of a contingency. Finally, utilities monitor the grid and operate grid equipment to ensure the balancing of load with generation supply in real time.

The starting point for the process is a load forecast for the customer base. Load must be matched with a set of units that economically satisfies the load. There are inherent limitations on grid capabilities that must be taken into account when setting a dispatch schedule. Similar to the planning of infrastructure additions, an integrated approach that considers generation resources as well as grid limitations must be applied to unit dispatch scheduling and operations. This integrated approach must address two key objectives: cost minimization and reliability.

Chapter 2 characterizes load requirements and presents the generation technologies that utilities own and dispatch. Chapter 3 presents the configuration, limitations, and operations of the grid that must be taken into account when setting a dispatch schedule. Chapter 4 discusses the day-ahead dispatch scheduling process.

1.1.3 Load and Supply Management: Midterm Planning

Midterm planning is a set of decisions that affects short-term dispatch scheduling but is distinct from short-term planning because the issues are addressed before the setting of the day-ahead dispatch. This set of decisions is also distinct from long-term planning because it involves management of existing supply resources as opposed to the addition of new resources. Midterm planning decisions include setting maintenance schedules for generation and transmission equipment, determining the best use of resources with limited availability (i.e., generation with limited use due to environmental regulations), and the pooling of resources with neighboring utilities. The objectives of midterm planning decisions are identical to those of long-term and short-term planning decisions: cost minimization and reliability.

As with short-term and long-term planning, an understanding of these issues requires knowledge of generation and transmission equipment, the subjects of Chapters 2 and 3. Midterm planning as its own topic is presented in Chapter 6.

Remark

- An additional activity that this text does not address is a utility's fuel procurement. This is an activity that is certainly within the power delivery chain; however, the text focuses on planning activities related to generation and grid infrastructure. Fuel delivery infrastructure and procurement is a separate topic that requires its own text.

1.2 THE POWER DELIVERY CHAIN IN A MARKET ENVIRONMENT

A market environment is one in which end customers choose their electric service provider from among several competing companies. This contrasts with a traditionally regulated environment in which a regulated utility is assigned a territory and services all customers within the territory. The market environment is referred to as a power market.

The power delivery chain in a market environment must address the three elements of the chain identified in Section 1.1. Within a market environment there is not a single agent that performs these tasks. Instead, many different market participants each perform specific functions. In this section, we identify the various market participants along with their market functions.

There are several functioning power markets throughout the world. Several markets exist in the eastern United States: PJM (Pennsylvania, New Jersey, and

Maryland), New England, and New York. A market exists in Texas, and California is moving toward the development of a market. Australia and New Zealand also operate power markets, and there are several European markets that are in various stages of development. Scandinavia has perhaps the world's most mature market. Every functioning power market in the world assigns a single regulated public agent to manage the grid because the transmission network must be managed and operated by a single agent that is indifferent to market outcomes. This agent is called the independent system operator (ISO). As such, there are no fully deregulated power markets. Instead, there is a degree of deregulation confined to some aspects of the electric power delivery chain.

As noted above, interdependence between grid and generation operations requires close coordination between these activities. Because there are no natural divisions between these activities, a distinguishing factor of different power markets is the role of the ISO in generation-related decisions. The ISO's role in the power delivery chain is presented below along with other market participants' roles.

The chapters corresponding to these activities are Chapters 7 through 10. Chapter 7 presents a market structure that accounts for operational requirements of the power delivery chain. Subsequent chapters describe the practice of market participants within the market structure.

1.2.1 Infrastructure Additions: Long-Term Planning and Investment Analysis

Long-term planning within a utility environment devolves into investment analysis in a market environment. Independent power producers (IPPs) are for-profit companies that construct, own, and operate power plants. The decision to construct a new facility is based on the economic analysis of an IPP along with its ability to obtain financing for the project.

Because of the interdependence between generation and the grid, a power plant cannot be constructed without consideration of its impact on the grid. Indeed, for the grid to accommodate a new power plant, additions to the transmission system must be considered. The ISO assumes the role of determining grid upgrades associated with a power plant addition. Financing, construction, and recovery of grid upgrades differ among the different markets.

Transmission equipment in a market environment is constructed and owned by transmission companies (TRANSCOs). Addressing infrastructure additions solely for reliability and grid security is an issue of paramount concern. Projects typically require combined regulatory and market-based mechanisms. The regulatory aspect tasks the ISO with proposing projects to a commission, as is the case with utilities. TRANSCOs should bid for construction and ownership rights and then lease the equipment to the ISO at agreed-upon terms.

Chapter 7 provides an overview of market structure that addresses investments, and Chapter 9 focuses on investment analysis.

1.2.2 Day-Ahead Scheduling and Real-Time Grid Management: Short-Term Planning and Asset Management

Short-term planning and real-time operations activities of utilities fall within the heading of asset management in a market environment. This is so because market participants refer to power plants and grid equipment as assets and it is in the short-term phase that assets are scheduled and dispatched.

Day-ahead scheduling proceeds through a market that the ISO operates. There are four market participants associated with these activities: the ISO, IPPs, retailers, and integrated energy companies. Retailers are companies that sign up end customers, procure electricity on their customers' behalf, and bill the customers. Retailers do not own and operate power plants; as noted above, IPPs perform this role. Integrated energy companies are those that provide the combined services of IPPs and retailers. Below, we do not refer to integrated energy companies as they assume the role of both IPP and retailer. Note that both retailers and IPPs require access to the grid; IPPs must be able to place their power on the grid, and retailers must be able to deliver power to their customers through the grid. It is the ISO's responsibility to ensure fair access to the grid.

Every day the ISO receives demand bids for electricity from retailers as well as supply bids of individual power plants from IPPs for the following day's scheduling. Using these bids, the ISO sets a dispatch schedule that awards the selected generation units a market clearing price that is commensurate with the bids of retailers. The ISO ensures that the transmission system is able to manage anticipated power flows as well as contingencies associated with the day-ahead schedule.

Real-time grid management, ensuring that power demand is equilibrated with supply while maintaining reliability and security standards in real time, is the responsibility of the ISO. The ISO has the authority to dispatch units in line with market awards, even though IPPs operate the power plants. The ISO also directs the use of grid equipment.

The contents of Chapters 2, 3, and 4 are critical to understanding asset management in a market setting; these chapters provide universal planning and operational principles that apply to power delivery. The structure of a market that accounts for operational issues identified in Chapters 2, 3, and 4 is presented in Chapter 7. Chapter 8 provides details of market participants' behavior in the short-term markets.

1.2.3 Load and Supply Management: Midterm Planning, and Risk Management

Midterm planning falls under the realm of risk management in a market environment. Risk management is the use of market instruments to align the financial positions inherent in a portfolio of end customers and assets with the company's risk preference. Adjustments to the market participant's portfolio are made through market sales and purchases that set the price of electric power well in advance of the delivery date for that power. Whereas the objective of a utility is to minimize the cost of service while maintaining reliability, the objective of a market participant is to capture profits while managing earnings risks. The market instruments available to IPPs and retailers are presented in Chapter 7. Limitations and use of the market instruments are presented in Chapter 10.

Remarks

- There is a natural relation between chapters devoted to utility planning and those devoted to the practice of market participants in a market structure. Chapter 8, *Asset Management in Short-Term Markets*, is the counterpart to Chapter 4, *Short-Term Utility Planning*. Chapter 9, *Investment Analysis: Long-Term Planning in a Market Environment*, is the counterpart to Chapter 5, *Long-Term Utility Planning*. Finally, Chapter 10, *Risk Management in the Midterm Markets*, is the counterpart to Chapter 6, *Midterm Utility Planning*.
- Stand-alone transmission projects in a market environment and the corresponding activities of TRANCOs are not addressed.

Chapter 2

Energy, Load, and Generation Technologies

A central concern of this book is the description of planning activities to ensure the balancing of generation supply and electricity demand in daily operations. To understand the process one must first be able to quantify demand and supply requirements. Additionally, knowledge of electricity demand characteristics and generation technologies provides a better understanding of planning activities. This chapter provides the fundamental information that underlies planning processes. Section 2.1 provides a basis for measuring energy requirements as well as converting requirements across different units of measurement. Section 2.2 presents the factors that influence demand. Section 2.3 presents the generation technologies that are most commonly used to provide electricity.

2.1 ENERGY, POWER, AND THEIR MEASUREMENTS

Energy comes in different forms. In electric generation there is conversion of energy from thermal energy to mechanical energy to electrical energy. This section discusses these forms of energy and power with three objectives. First, we wish to create a level of familiarity with the units and demonstrate the conversion of units between the energy types. Second, we provide a brief description of the conversion process. Third, we wish to provide an understanding of the output ratings for generation units, a unit's capacity. An understanding of the difference between power and energy is important for all these objectives. Accordingly, the section highlights this distinction.

Mechanical energy is presented first, as it is the most intuitive. Afterwards we explain the basics of electric generation by describing the conversion of mechanical to electric energy. This is done through an analogy with mechanical energy. Then units of electric energy are presented, along with an analogy between these units and those presented for mechanical energy. Section 2.1 concludes with the definition of the standard energy measurement for fuels and an example of converting energy

Electric Power Planning for Regulated and Deregulated Markets. By Arthur Mazer

units. The British measurement system is adopted here to comport with utility standards.

2.1.1 Mechanical Energy, Power, and a Mechanical Circuit

Mechanical energy is required to move physical objects that resist motion. A unit that is often used to measure mechanical energy is the foot-pound. Suppose that a body resists motion with 1 pound of force. Then the amount of energy required to move the body through 1 foot is a single foot-pound. A weightlifter lifting 400 pounds from the ground to a distance of 7 feet above the ground must impart 2800 foot-pounds of energy into the weight.

The weightlifter does not create energy to lift the weight. Instead, he converts energy through oxidation. Fuel from food is stored within the body. The energy in this fuel is oxidized and eventually transferred to the weight. The rate at which energy transfer occurs is called power. One way of measuring power is in units of foot-pounds per second, indicating the amount of energy that can be transferred per second.

When planning tasks that require energy, stipulation of the energy requirement is insufficient. One normally wants to accomplish the task within a time frame; accordingly, power requirements must also be established. In the above example, it is possible for a small child to lift 400 pounds of weight up to a height of 7 feet. The child could not do this all in one shot; the weight would have to be broken down into smaller weights that the child could carry. The child could then carry these smaller weights up a series of stairs to a height of 7 feet and deposit them at the top. After many minutes, the child, just like the weightlifter, will have imparted 2800 foot-pounds of energy into the weights. Of course, since the child does not possess the power of the weightlifter, the child cannot lift the weight as quickly. As noted above, if one wants to lift the weight within a certain time frame, the power required to do so must be established.

The concepts of power and energy apply to electricity as well. The next section discusses the conversion of energy from mechanical to electrical. To assist with that discussion, we first present a mechanical example that provides analogies with quantification of energy and power in electrical systems.

EXAMPLE *A Mechanical Circuit*

When a body is moved from a lower height to a higher height, the energy required to elevate the body is stored within the body as potential energy. With the use of mechanical devices, this energy may be harnessed.

Suppose that we have a series of bowling balls weighing 10 pounds each that we are raising from the ground to a platform 20 feet high. A machine pushes the bowling balls up a slide. Suppose we want to keep the bowling balls moving at a rate of 5 bowling balls per second. We calculate the power requirement for this system as follows:

$$\text{PowerRequirement} = 20\text{ ft} \times 50\frac{\text{lb}}{\text{s}}$$
$$= 1000\frac{\text{ft-lb}}{\text{s}}$$

From the power it is possible to determine the energy consumption after a given time frame. The amount of energy consumed after 30 seconds is calculated as follows:

$$\text{EnergyConsumption} = 1000\frac{\text{ft-lb}}{\text{s}} \times 30\text{s}$$
$$= 30{,}000\text{ ft-lb}$$

The above two equations are useful for understanding power and energy units used in electric systems.

It is possible to create a complete circuit by incorporating a mechanism that returns the bowling balls to their initial position at ground level. For example, a Ferris wheel-type device could guide the bowling balls back to their initial position. Balls would be loaded onto the Ferris wheel at the top of the slide and unloaded at the bottom. This would rotate the Ferris wheel, and we could connect the Ferris wheel to a shaft that would turn a machine, for example, a water pump. If we do this just right, then the flow rate of the bowling balls remains constant along its entire path. The cycle repeats itself.

2.1.2 Electric Generation and Units of Measurement

This section describes a generator as well the process of generating electricity in a circuit. Practical design and planning of power systems requires quantification of production to match needs, model and control systems, and appropriately engineer a generator. After describing electric generation, the section provides definitions of electrical energy and power measurements that are used throughout the remainder of the text.

Figure 2.1 provides a diagram of a generator. The components include an electromagnet that is connected to a rotating shaft and coils made of conducting material that can be connected and decoupled to a grid. The rotating shaft on which the electromagnet rotates is called the rotor. The coils are called stators. In the diagram there is a single stator made up of two connected coils.

An electric current is the drift of electrons in a common direction. On their own, electrons do not drift through the wire in a common direction; this requires the introduction of an electrical force on the electrons. Rotating the magnet causes a fluctuation in the magnetic field through the stators. The fluctuating magnetic field induces an electric force on the electrons that causes them to drift as a current. Once the electrons are drifting in a common direction, energy may be extracted to run electrical appliances.

Mechanical energy is required to keep the rotor spinning. (If the rotor doesn't spin, the magnetic field does not fluctuate and no electric force is applied to the electrons in the circuit.) The current through the stators generates its own magnetic force that operates on the rotor's electromagnet in the opposite direction of the

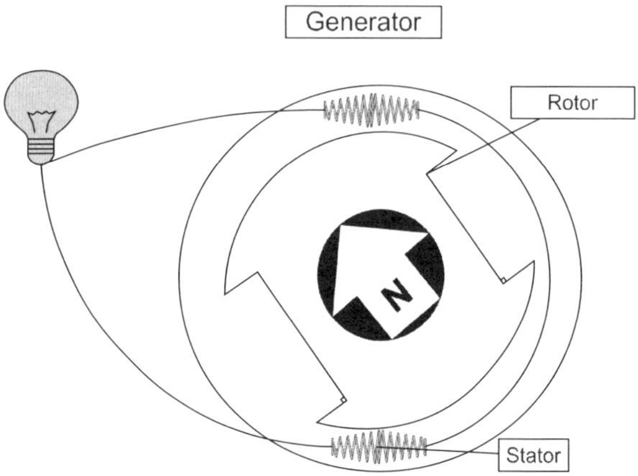

Figure 2.1

rotor's spin. This is somewhat analogous to the mechanical principle that if body A pushes on body B, then body B pushes back on body A with an equal and opposite force. The countervailing magnetic force acting on the rotor must be overcome to keep the rotor spinning. In actuality the magnet is kept spinning by a turbine that rotates the shaft, but imagine that the magnet is kept spinning by a force applied to the magnet's tip. The mechanical energy required to keep the magnet spinning is this force times the distance that the magnet moves as it goes around its circular path. The power that is input into the system is the force times the speed of the magnet tip.

Remarks

- The generator depicted in Figure 2.1 is a simplification of utility generators. Utility generators have three sets of stators. Each stator connects to the transmission system. A rotor can have more than a single magnet. It can have multiple magnets with multiple sets of poles. Aside from the rotor and stators, an additional element is a cooling mechanism. There is significant heat development in the rotor and stators during generation. Without a cooling process, the heat buildup would liquefy or burn the rotor and stators. While rotor temperatures vary with equipment and operations, typically cooling mechanisms maintain rotor temperatures of 240°F. The most common coolant is liquid hydrogen.
- A rotor with a single set of poles rotates at very high speeds, 60 rotations per second. For larger generators, the rotor radius is 2 feet. At 60 rotations per second, the tip of a rotor moves at a speed of 770 ft/s, which is 514 miles per hour.

To introduce energy and power measurement, analogies will be drawn with the example from Section 2.1.1, the mechanical circuit. In that example, a machine lifts bowling balls at a measurable flow rate up a slide to a specified height. The balls

Table 2.1 Mechanical and Electric Circuit Analogy

Property	Mechanical circuit example	Electric circuit example
Force creation	Gravitational force of earth	Electric force from rotating magnet
Energy carrier	Bowling balls	Electrons
Energy producer	Machine pumping balls	Machine rotating generator
Energy consumer	Water pump	Electric motor
Energy levels	Height	Electric potential level
Flow rate (current)	Flow rate of bowling balls	Flow rate of electrons
Power	Flow rate times height change	Flow rate times change in electric potential
Energy input	Power times time	Power times time

then provide power to a water pump and return to their initial starting point via a Ferris wheel. The generator analogy is that the generator pumps electrons at a measurable flow rate through a circuit from a lower to a higher electrical energy level, increasing their potential energy. The electrons then flow through the circuit from the higher energy level toward the lower energy level as a current. As the electrons flow along the circuit they may be used to power devices such as electric motors. Electrons at the terminal point of the circuit (same as starting point) all have the same energy level. Table 2.1 summarizes this analogy.

Standard units are necessary to measure current, energy, and power. We present the standard industry units.

Current: Ampere

The unit for current is the ampere. This provides the flow rate of electrons that are flowing through a cross section of the circuit.

Electric Potential Level: Volt

The electric potential level of a point on a circuit (analogous to height in the mechanical example) is measured in volts. Electrons at the same voltage level have the same potential energy. Within a circuit, cross sections that are transverse to the direction of the current have nearly the same voltage.

Power: Megawatt

Utilities most often measure power in megawatts (MW). In the mechanical example the power input is determined by multiplying the change in height, 20 feet, by the flow rate of the bowling balls. Similarly, one attains the power input in an electrical circuit by multiplying the change in voltage around the circuit by the current (flow rate). Accordingly the unit volt-ampere is a power unit. One volt-ampere is the amount of power associated with moving a current with a flow rate of 1 ampere through a change of electrical potential level of 1 volt. One million volt-amperes are equal to 1 MW.

Energy: Megawatt-Hour

Recall the mechanical example, in which the energy input over a specified time duration is determined by multiplying the power input by the time duration. To determine the energy consumed in a circuit, one applies the same principle. This provides the standard unit of measurement for electric utilities, the megawatt-hour (MWH). One MWH is the amount of energy produced by a unit generating 1 MW of power over 1 hour.

The size of a generation unit is measured by its power rating and is called the plant's capacity. By the definitions above, a plant with a capacity of 400 MW is able to generate 400 MWH of energy in 1 hour.

2.1.3 Fuels and Conversion of Units

The above section explains the conversion of mechanical to electrical energy. The starting point in most power generation is energy released through the combustion of fuel. Potential energy is stored as fuel and released in the combustion process. The industry standard for measuring the amount of energy stored in a fuel is the MMBTU. A single BTU is the energy required to increase the temperature of 1 pound of water 1 degree Fahrenheit. An MMBTU is 1 million BTUs.

In a plant, the physical conversion of MMBTU to MWH determines the plant's efficiency. A single MWH is equivalent to 3.412 MMBTU. The conversion rate of MMBTU to MWH provides a plant's efficiency and is known as the plant's heat rate. Accordingly, if a generator could convert all of the energy from the fuel into electric energy, its heat rate would be 3.412 MMBTU/MWH. Generators cannot come anywhere close to 100 percent efficiency. The better units are around 50 percent efficient. This translates into a heat rate of 6.824 MMBTU/MWH, meaning that producing a single MWH of electric energy requires burning fuel with a heat content of 6.824 MMBTU.

Section 2.3 provides more detail on the physical conversion process from thermal to mechanical energy. We conclude Section 2.1 with examples of energy and power measurement conversions followed by a qualitative section summary. Table 2.2 is provided to assist with the conversion of measurements.

EXAMPLE ***How Many Light Bulbs from a Power Plant?***

A modern-generation turbine can have a capacity of 1000 MW; it is capable of providing 1000 MW of power. We use Table 2.2 to illustrate just how much power this is. Consider how many light bulbs could be lit with 1000 MW of power. A typical light bulb in a reading lamp requires 100 W of power. From Table 2.2 it is seen that a 1000-MW unit is capable of delivering enough power to light 10 million light bulbs.

Table 2.2 Energy and Power Unit Conversions

	Energy					
	KWH	MWH	Watt-Hour	BTU	MMBTU	ft-lb
KWH	1	0.001	1,000	3,412	0.003412	2,655,224
MWH	1,000	1	1,000,000	3,412,141	3.412	2,655,223,832
Watt-hour	0.001	0.000001	1	3.412	0.0000034	2,655
BTU	0.000293083	—	0.2931	1	1,000,000	778
MMBTU	293	0.293	294,118	0.000001	1	0.000778
Ft-lb	—	—	0.0003766	0.001285	1285	1
			Power			
	KW	MW	Watt	BTU/hour	MMBTU/hour	ft-lb/s
KW	1	0.001	1,000	3,412	0.003412	738
MW	1,000	1	1,000,000	3,412,141	3.412	737,562
Watt	0.001	0.000001	1	3.412	0.0000034	0.7375
BTU/hour	0.000293083	—	0.2931	1	1,000,000	0.2161
MMBTU/hour	293	0.293	294,118	0.000001	1	—
Ft-lb/s	—	—	0.0003766	0.001285	1285	0.0003

An empty entry denotes less than 1 per 10 million.

EXAMPLE ***Truck Toss***

Another way of viewing the power behind 1000 MW is to note that 1 MW is equivalent to 737,562 ft-lb/s and 1000 MW of power is equivalent to 737,562,000 ftlb/s. To place this in context, consider the following scenario. A typical pickup truck is on the order of 5000 pounds. Suppose we utilize a generator's 1000 MW of power to lift the pickup truck straight up over a 1-sec time interval and then release the vehicle. The generator imparts 737,562,000 ft-lb of energy into the truck at the time of the truck's release. After release, the vehicle will continue on its upward path and reach a height of 147,493 feet (147,493 feet = 737,562,000 ft-lb/5000 lb), nearly 28 miles, before falling back down to Earth.

Summary

- Energy comes in various forms and can be converted from one form to another. Most electric generation processes burn fuel to create thermal energy. This energy is converted to mechanical energy. The mechanical energy is then converted to electrical energy.
- The common unit for measuring the content of thermal energy in fuels is the MMBTU, which amounts to 1 million British thermal units (BTUs). A single BTU is the amount of energy required to raise the temperature of 1 pound of water by 1 degree Fahrenheit.
- The common unit for measuring electrical energy is the megawatt hour (MWH). One MWH is equivalent to 3.412 MMBTU.
- Power is the rate at which one can deliver energy. Electric power is measured in megawatts (MW). The power rating of a generation unit is called the unit's capacity. A 400MW generation unit can deliver 400 MWH of energy in 1 hour.
- A measurement used to determine the required MMBTU thermal energy input required to output a MWH of electrical energy is the heat rate. The units of the heat rate are MMBTU per MWH. To convert the heat rate to efficiency, divide 3.412 by the heat rate.

2.2 LOAD

Load refers to customer demand for electricity. This section categorizes the different load types: residential, commercial, and industrial. The discussion first focuses on the major load drivers: population, gross domestic product (GDP), temporal behavior patterns, and various weather parameters. Additional characteristics of load that are required in the planning process are then presented.

2.2.1 Customer Classification and Load Drivers

Residential Load

The primary consumption of residential customers goes toward lighting and running appliances. Appliances that require high power include refrigeration and air conditioning. Table 2.3 presents power requirements for various appliances. Summing up the requirements provides a large overall requirement per household.

Table 2.3 Power Requirements of Household Items

Appliance	Power requirement in watts (while in operation)
Window air conditioner	600–3500
Refrigerator	400–700
Freezer	500–800
Television	300
Computer (CPU)	150
Vacuum cleaner	500–1500
Stove	1500 (1 heating element)
Oven	3400
Microwave	600–1000
Washer	1200
Dryer	5400
Dishwasher	1500
Light bulb	50–150
Hair dryer	1200
Water heater	4000

Although the power requirement of a given household at a particular time may be quite high, the average household requirement is approximately 1800 watts. The figure is far lower than the sum of individual appliances because households do not run all their appliances simultaneously. Additionally, appliance use varies among households: Not everyone runs their clothes dryer at the same time. There are exceptions: Air conditioning usage is highly correlated across households since households respond to the same weather patterns. Weather is one among several factors that determine load. A brief discussion of load drivers and their impact follows.

The critical load drivers are listed and discussed below.

Population Although energy consumption varies from household to household, overall residential load is highly correlated with population. It stands to reason that more households create more demand.

GDP The wealth of households is also highly correlated with demand. Wealthier households have central air conditioning, larger living space, larger refrigerators, and more appliances. They are less sensitive to utility bills and accordingly less likely to economize their discretionary consumption, for example, turning up the thermostat on central air conditioning. In economic upturns, residential demand per household increases.

Temporal Behavior Patterns Electric consumption varies over time cycles in relation to variation in human activities. The cycles are associated with different timescales, seasonal, weekly, and daily. A modest seasonal consumption cycle follows the vacation and school cycles. Weekly cycles that follow a weekday and weekend pattern are very apparent. There is also a noticeable daily consumption pattern that follows early morning hours, work hours, evening hours, and sleeping hours. The temporal behavior patterns are also very much linked to weather patterns, as discussed below.

Weather Parameters Weather has the largest impact on load. Weather follows seasonal and daily patterns, which coupled with temporal behavior impact residential load. Below we indicate the more significant weather parameters and discuss their impact on load.

- Temperature: Residential load responds to temperature more than any other weather variable. Air conditioning demand increases in response to high temperatures. Heating demand increases in response to low temperatures. Demands on refrigeration can fluctuate in response to changes in indoor temperatures.
- Humidity: Air conditioning demand increases with increasing humidity.
- Recent temperature history: Heating and air conditioning demand increase as intemperate periods persist for longer durations. This is caused by air pockets dissipating through insulating materials and external walls and decreasing the effectiveness of the insulation.
- Sunshine and cloudiness: Cloudiness increases daytime light usage. However, it decreases air conditioning usage as less radiant energy heats up rooftops. The net effect is either positive or negative depending on lighting and air conditioning requirements. In Belgium, load decreases on rare sunny days as the reduction in lighting need is greater than the increase in air conditioning use. The opposite occurs in Tucson.
- Declination of the sun: The rate at which solar radiation heats up a roof or exterior walls depends on the declination of the sun. Higher heat transfer increases air conditioning load while decreasing heating load.
- Wind: Increased wind speed decreases the effectiveness of insulation. Additionally, external air may enter a house through gaps in windows and doors at higher wind speeds. Accordingly, wind increases residential air conditioning and heating demand.

Remark

- Residential load requirements may have peculiarities in certain geographic locations. For example, areas of Florida have a large number of heated swimming pools. When a cold front hits Florida, residential demand spikes significantly and can match the summertime air conditioning load.

Commercial and Industrial Load

Commercial load arises from commercial activities other than manufacturing; predominantly the service sector of the economy. By far the greatest power usage in the commercial sector is indoor temperature control and lighting. Industrial load results from electricity requirements of manufacturing processes. Load drivers affect commercial and industrial load as follows.

Population Commercial and industrial activity increase with increasing population. Accordingly, as with residential load, commercial and industrial load increases with increasing population.

GDP Commercial and industrial load is increased or decreased by strong or adverse economic growth. This follows from an overall increase or decrease in commercial and industrial activity associated with corresponding strengthening or weakening of the GDP.

Temporal Behavior Patterns As with residential load, commercial activity has seasonal, daily, and hourly patterns that affect load. These patterns are highly predictable as commercial operations work on a strict time clock.

Manufacturing process follow strict and planned timelines. Industrial load follows the processing timelines and is very predictable.

Weather Parameters As indoor temperature control is a sizable component of commercial load, weather impacts are similar to those on residential load. However, commercial buildings are generally better constructed than residential buildings; accordingly, they are not as sensitive to weather changes as residential buildings.

With the exception of extreme weather such as hurricanes, industrial load is not sensitive to weather conditions.

2.2.2 Further Characteristics of Load and Useful Definitions

This section presents additional characteristics of load that are important in power operations and planning.

Load Shape

A load shape is the power requirement as a function of time over a given time interval. There are daily, weekly, and seasonal load patterns. Loads change over the course of the day because of behavioral patterns and in response to changing weather conditions. Figure 2.2 provides graphs of daily load shapes for Eastern PJM, an area including Pennsylvania, New Jersey, and Maryland, over a 24-hour time frame. The horizontal axis provides the time, and the vertical axis provides the power requirement in MW. (Note that the total energy requirement in MWH is given by the area under the curve.) Graphs are plotted for January 13, 2005 and July 27, 2005.

The distinguishing feature of the summertime daily shape is the magnitude of the change in power requirements over the day. The change ranges from a minimum

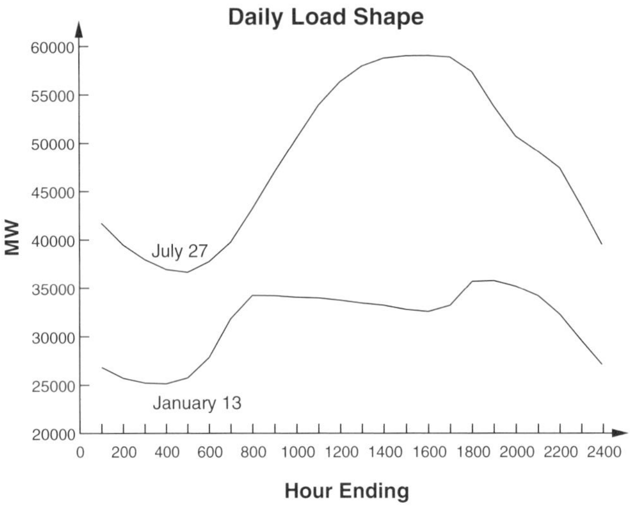

Figure 2.2

requirement of 36,667 MW at the hour ending at 0500 to a maximum requirement of 59,031 at the hour ending at 1600. The magnitude of the difference is 22,364 MW. As noted above, the fluctuation in load requirement is due to a combination of behavioral and weather-related load drivers: Residents arrive home and respond to high daily temperatures with air conditioning demand to create the peak. Cooler weather and a sleeping population create the least demand.

An interesting feature of the wintertime daily load shape is that the shape displays a double peak, one at the hour ending at 1100 and one at the hour ending at 1900. Both peaks are attributable to a rise in heating demand. The overall demand fluctuation for January is much smaller than that for July. The minimum demand occurs at the hour ending at 0400, with a value of 25,144 MW, while the maximum demand is 35,785 MW, occurring at the hour ending at 1900. Whereas the difference between minimum and maximum for the summertime day is 22,364 MW, the difference between minimum and maximum for the wintertime day is only 10,641 MW. The daily load shape influences the type of units that are constructed as well as the daily dispatch of those units. More on the dispatch of units is presented in Chapter 4, *Short-Term Utility Planning*.

Within the power industry there are time blocks associated with load levels. The naming and hours associated with these time blocks are motivated by the daily load shapes. Table 2.4 presents the time blocks along with their associated times. Be aware that the conventions are not universal and locations may have their own standards.

Weekly load shapes arise because of differences in weekday and weekend activities. Figure 2.3 provides a graph of a week's power demand in PJM. The overall demand during the weekend period is significantly lower than that of the weekday period. The weekly load shape influences the maintenance scheduling of power

Table 2.4 Time Blocks

Name	Hours
Weekday on peak	Weekdays from 7 A.M. to 11 P.M.
Weekday super peak	Most useful during summertime peaks. 12 noon to 8 P.M.
Weekday off peak	All weekday hours that are not on peak
Weekend and holiday on peak	7 A.M. to 11 P.M. on weekends and holidays
Weekend and holiday off peak	All weekend and holiday hours that are not on peak
Shoulder hours	Hours that bridge on peak and off peak periods. 5 A.M. to 8 A.M. and 10 P.M. to midnight

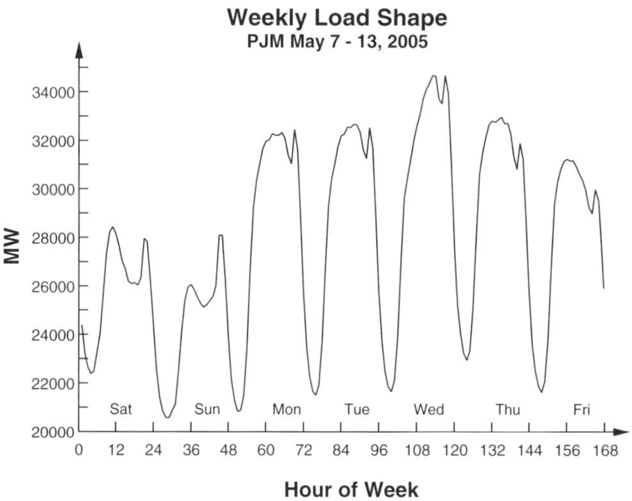

Figure 2.3

plants, when possible, maintenance is scheduled over weekends so that plants are available to satisfy weekday load.

Aside from the seasonal effect on the daily load profiles there is also a seasonal effect on overall demand. Figure 2.4 provides a graph of the load in MW over a year's time for the same location and year as the previous graphs, Eastern PJM 2005. The horizontal axis provides the hour of the year, and the vertical axis provides the MW load for the given hour. There is an upper envelope corresponding to the daily maximum load and a lower envelope corresponding to the daily minimum load. The upper envelope is at its highest levels in the summer and at its lowest levels during the fall and spring. Accordingly, fall and spring are typically the seasons when maintenance for power plants requiring significant overhauls is scheduled. More on maintenance scheduling is provided in Chapter 6, *Midterm Utility Planning*.

A point to note is the width of the lower and upper envelopes of the points. The envelope provides the seasonal range of requirements. The envelope width increases

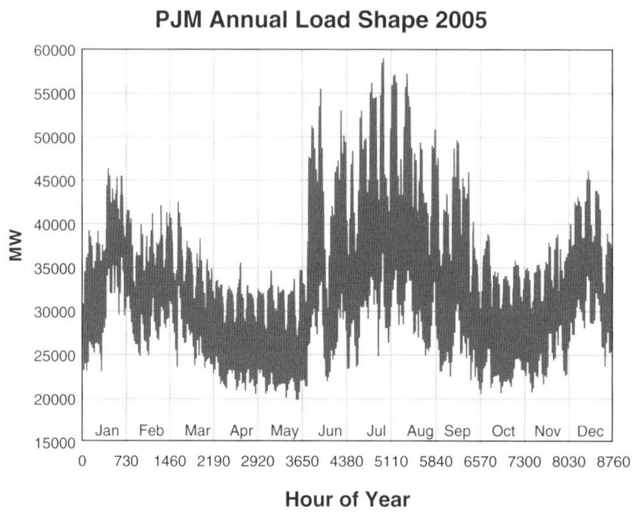

Figure 2.4

substantially during the summer months. The size and shape of the upper envelope influence the number of power plants (quantity of capacity) as well as the technology of the power plants that utilities select. Providing sufficient power to meet summertime loads requires carrying a surplus of generation plants for the remainder of the year. The surplus plants are typically less efficient than the generation plants that operate throughout the year. However, the surplus plants are also less expensive to construct. More details of technology selection are presented in Chapter 5, *Long-Term Utility Planning*.

Interruptible Load

Utilities offer customers reduced prices in return for the right to interrupt service. This provides the utility with the ability to reduce load when there is a shortage of generation capacity. The price for interruptible supply depends on the number of interruptible hours per year that the customer accepts: A greater number of interruptible hours provides more favorable rates to the customer.

Remarks

- Load predictions are made over different time horizons. Short-term load predictions are required for scheduling a next-day dispatch. Long-term load forecasts are required for planning retirements and additions.
- Short term-load forecasts, day ahead, do not depend on population and GDP. Short-term load forecasts are sensitive to temporal behavior and weather. Forecasts of temporal behavior are fairly accurate. However, weather forecasts are less certain. Accordingly, weather is the greatest uncertainty in short-term load forecasts.

- As residential load is the load component that is most sensitive to weather, residential load is also the least predictable over short time horizons. Load forecasting is discussed further in Chapters 4 and 5.
- The main driver of seasonal load fluctuations is weather. Residential load is the load component that is most sensitive to weather and accordingly the component responsible for fluctuations. As will be seen in subsequent chapters, load shapes with significant peaks and valleys over daily and seasonal ranges are less economical to serve than flat load shapes.
- Long-term weather forecasts are not available. Accordingly, long-term load forecasts utilize historical weather and are influenced by GDP and population outlooks.

2.3 GENERATION TECHNOLOGIES

Central to all planning activities and operations is an understanding of the generation technologies available for servicing load. In this section we describe the most common generation technologies. Sections 2.3.1 describes the most common form of generation, steam generation. Sections 2.3.2 through 2.3.7 present the specifics of conventional technologies: coal, nuclear, combustion turbines, combined cycle gas turbines, hydro systems, and wind generation. The discussion focuses on the thermomechanical systems, used to spin the generator, maintenance, and operations. There is also a discussion of environmental impact and mitigations. Section 2.3.8 concludes the chapter with a summary that presents advantages and disadvantages of different technologies.

2.3.1 Steam Generation: System Components, Efficiencies, Operations and Maintenance

Steam generation is a common form of generation using different fuel types: coal, nuclear fuel, gas, oil, diesel, waste, and others. This section describes the components that are common to all steam generators and the function of these components. The section then discusses plant efficiency, operations, and maintenance.

Figure 2.5 provides a schematic of (a) coal steam power plant. The plant consists of a burner, boiler, tunnel, turbine, condenser, and return channel. Thermal energy is converted to mechanical energy through a steam cycle. The cycle follows a sealed pathway along points that are enumerated below.

1. Water is heated in the boiler to create high-pressure steam.
2. The steam exits the boiler and proceeds through a tunnel directed at a turbine. The steam follows a pressure gradient, moving toward low pressure.
3. The steam flows through the turbine, spinning the turbine. The turbine is connected to a shaft that spins the generator.

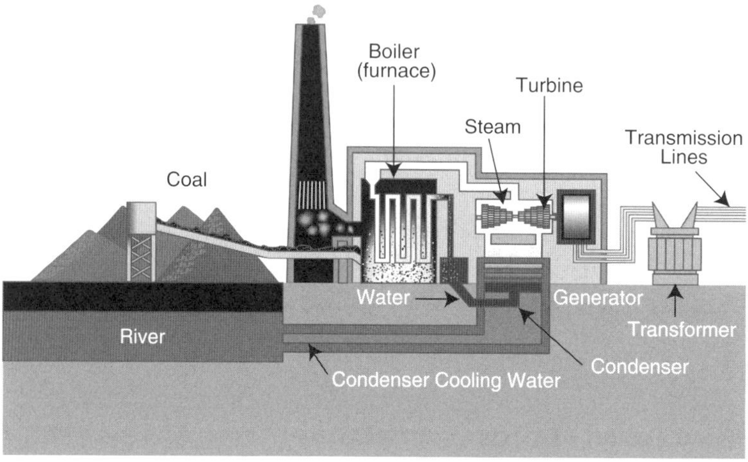

Figure 2.5

4. The steam is channeled toward a condenser, where it condenses to water. This creates low pressure at the condenser. The condenser operates through a heat transfer mechanism. A coolant (cold water) is pumped from a source to a heat exchange facility that is in contact with the steam inside the condenser. Heat is transferred from the steam to the coolant, and the steam condenses. The coolant is returned to its source, usually a nearby river. The coolant circulation system is independent from the circulatory fluid that turns the turbine, that is, the coolant and the steam do not mix.

   In many cases, very large units require cooling towers. The function of a cooling tower is to release the waste heat that has been transferred to the coolant into the atmosphere rather than discharging the waste heat into a nearby river. This is done so as not to disturb the ecosystem of the river.

5. The condensed water is pumped back to the boiler through the return channel, where the cycle is repeated.

Efficiency

Steam plants operate at different efficiencies depending on technology and environmental factors. However, a typical coal plant converts between 30 percent and 35 percent of the thermal energy to electric energy. Although losses depend on the specifics of the unit as well as external operating conditions, the following provides a typical profile of loss accumulations.

- Boiler losses. Boiler losses occur because not all thermal energy from the fuel source is transferred, to the water and steam. Around 85 percent is transferred while the rest is convected out the chimney.
- Phase transition (water to vapor). Phase transitions from liquid to gas require energy without increasing temperature. Water changes from a liquid to a gas

at approximately 212 degrees F, requiring energy of 900 BTUs/lb. This energy is lost as it cannot be called upon to do work. The phase shifting energy is 50 percent of the energy transferred to the circulating fluid. Accordingly, the energy available for conversion to mechanical energy is also 50 percent of the transferred thermal energy.

- Turbine efficiency. The turbine converts the mechanical energy of the moving steam into rotational energy, rotating the generator. The turbine is not able to convert all the energy because some remains with the steam. The converted energy is around 82 percent of the available energy.
- Plant operations. Operating plant systems such as pumps and the electromagnet on the rotor require energy and can be considered as system losses. These losses are negligible in comparison with those mentioned above. Also, generator losses during the conversion of mechanical to electric energy are negligible.

One arrives at the total efficiency by multiplying the efficiencies along the circulatory pathway.

$$\text{Total efficiency} = \text{eff}_{\text{boiler}} \times \text{eff}_{\text{phase transition}} \times \text{eff}_{\text{turbine}}$$
$$= .85 \times .5 \times .8 = .32$$

The heat rate associated with an efficiency of .32 is 10.7 MMBTU/MWH.

Remark

- Power plant efficiencies change with weather conditions. This is predominantly due to the effectiveness of the condenser under different weather conditions. The condenser is more effective in cold weather. The cold weather enhances the development of the low-pressure zone within the condenser, and the rate of steam flow across the turbine increases.

Operating Steam Plants

Below we describe common operational procedures for steam plants. Operational features that are specific to types of plants are addressed in their independent sections. The operations are discussed within the framework of a cycle: starting a plant, running it at optimal capacity, ramping down the plant, and shutting it off.

We assume that in the initial state the plant is idle. It is not electrically connected to its outgoing transmission lines, and there is no fuel burn. There is little vapor in the system; most of the fluid is liquid. There is no pressure differential between the boiler and the condenser; accordingly, there is no fluid circulation, and the turbine is at rest. The pump circulating liquid water from the condenser to the boiler is also not operating. The operational steps are as follows.

Step 1. Heat the Boiler The first step in starting a generator is to raise the temperature of the water in the boiler to create a critical level of steam pressure required to rotate the turbine. The time requirement for doing this varies by system and the temperature state of the system at start-up. If the system has been at rest for many hours and the components have all reached ambient air temperature, it will take a relatively long time to reach critical temperature. This is because the boiler tank and all the piping in the boiler tank in addition to the water must be heated. Naturally, the hotter the initial temperature of the system is, the sooner one can reach the critical level of steam pressure required to rotate the turbine. Many plants have an auxiliary gas-fired heating system to keep the boiler warm when the plant is not in operation.

Step 2. Turn On Pumping Systems: Condenser and Return Channel The coolant pumping system pumps the coolant into and out from the condenser. Another pumping system circulates the condensed water from the condenser to the boiler. These are switched on at the beginning of the start-up sequence.

Step 3. Circulate the Generator Coolant There is a separate cooling system to keep the electric generator from overheating. Before generating electricity, the generator coolant must be circulated. Otherwise, the generator will be damaged irreparably.

Step 4. Activate the Exciter The exciter is a device that controls the strength of the magnetic field emanating from the rotor's electromagnet. Recall that the electromagnet on the rotor must produce a magnetic field in order for the generator to generate electricity. The power output from the generator is proportional to the strength of the electromagnet's magnetic field. This is an important control feature that is further discussed in Step 8 and in Chapter 3.

Step 5. Synchronize Output with the Grid Initially, a generator's stators are decoupled from their outgoing transmission lines. We say that the generator is offline or decoupled from the grid. Before coupling with the grid, the generator's output must be synchronized with that of the transmission system and stabilized. We call this synchronizing with the grid. Synchronizing with the grid requires that the generator's rotational speed and voltage output are coordinated with those of the grid. Further properties of synchronization are discussed in Chapter 3, where alternating current is presented.

Step 6. Connect to the Grid Before coupling to the grid there is no power output from the generator. There is a voltage potential difference on the stators. Connecting to the grid requires closing switches that physically complete the circuit between the stators and transmission lines. Once the circuit closes, the generator is producing power and current is flowing to customers.

Step 7. Bring the Unit to Minimum Power Threshold Once coupled to the grid, the generator produces power. At low power output turbine rotation is mechanically unstable and excessive vibrations of the generator and turbine are evident. There is a threshold known as the minimum power output where the vibrations cease. To avoid unnecessary mechanical stress, a unit must be brought to an operational level at or above its minimum power threshold.

Step 8. Control Output Control of power output is accomplished by manipulating the exciter (this controls the strength of the magnetic field emanating from the rotor's electromagnet) and controlling the fuel burn rate. Increasing power output requires an increase in magnetic field strength as well as the burn rate of the fuel. There are limitations on the speed with which one can increase and decrease power. These limitations are known as the ramp-up and ramp-down rates.

Each generator has an optimal power output level as well as a maximum power output level, the capacity. The optimal power output level is the point at which the conversion of fuel to power is most efficient. The optimal power output level is close to but often does not coincide with the capacity.

With the controls discussed above the plant can be driven to and maintained near its optimal power level. Alternatively, a unit may be controlled to provide fluctuating power output in response to fluctuating load requirements. Further details of controlling output are provided in Chapter 3.

Step 9. Bring the Unit Offline and to a Rest State Bringing a unit back to rest requires reducing output to zero by reducing fuel burn in order to bring the turbine to rest while simultaneously reducing the strength of the electromagnet's magnetic field by adjusting the exciter. The stators are then decoupled from the grid, and the remaining systems are switched off.

Note that in going from a rest state to a state in which the unit outputs to the grid requires alteration of the physical properties of the unit. The increase in temperature of the boiler, tubing, and turbine creates stress and metal fatigue, particularly in the turbine. Maintenance implications are discussed below.

Outages and Maintenance

An unfortunate tendency of equipment is that it breaks down. Scheduled outages, also referred to as planned outages, occur at regular intervals for performance of routine maintenance that reduces the likelihood of equipment failure. Routine maintenance consists of replacing system components in line with manufacturer specifications, replenishing lubricants, and routine inspections to determine the condition of system components. Units are not able to operate during planned outages.

Aside from planned outages, unforeseen equipment failures cause operators to take units offline. Such outages are referred to as forced outages. For steam plants, the most common cause of forced outages is steam leaking from perforated pipes within the boiler. Leaks can be serious, requiring the unit to be taken out of service, or not so serious. If the leak is not serious, a unit can continue operating during periods of high load and maintenance can be delayed until the weekend.

The most serious problems causing forced outages are due to turbine malfunctions. The turbine is the workhorse of a generation plant. The turbine converts the steam flow into rotational mechanical energy that is used to rotate the generator.

There is a metal encasement that envelopes the turbine. This encasement is sealed to ensure that the steam remains within its prescribed pathway. There are inlet and outlet sectors leading into and away from the turbine. These sectors must be coupled to the entrance and exit steam pathways. Between the inlet and outlet is a series of rotating and fixed blades arranged in rows. All rows of rotating blades are mounted to a single rotating shaft that spins the generator. The motion of the steam imparts a force on the blades that causes the blades to rotate. Between the rotating blades are fixed blades. Their function is to maintain an optimal flow pattern for the steam. To ensure that steam does not exit the flow path through the rotating shaft, this shaft must be sealed. Additionally, there is a control valve at the inlet sector. The control valve provides control over the steam inlet, allowing for mechanical control of power output and turbine speed. In emergencies this can be used to quickly take a unit offline.

The turbine is divided into different sectors, a high-pressure sector, a medium-pressure sector, and a low-pressure sector. The high-pressure sector is located toward the inlet, and the low-pressure sector is located toward the outlet. The blade sizes within these sectors vary. Blades within the high-steam pressure sector are smaller than those in the low-steam pressure sector.

Turbine malfunctions become apparent through excessive vibrations and performance reductions. Turbine vibrations are monitored by engineering staff. Vibrations beyond operational tolerances require immediate unit shutdown.

Cycling a unit off and on increases strain on the turbine in two ways. Thermal stress from the heating and cooling of the turbine causes cracks in the turbine casing and blades. Additionally, cycling of a unit also increases mechanical stress on the turbine as the generator is brought to its minimum power. Mechanical stress evidenced by vibrations during the start-up phase also fatigues the sealed bearings that encase the rotating shaft at the point where the shaft exits the turbine.

Outages due to turbine problems typically last longer than other types of outages. On detection of a problem, a decision must be made to perform a visual inspection or to do further diagnostic testing. A visual inspection requires opening the external casting that envelopes the turbine. If this is necessary, the unit will be out for a long time, perhaps a month. The process will require opening the casting, inspecting the entire turbine, performing any repairs or maintenance, cleaning deposition, and closing and resealing the casting.

Diagnostic testing involves running the turbine at specified temperature and pressure levels and examining the output as well as measuring vibrations. The output and vibrations should be at prescribed levels. Modern machines have continuous monitoring capability. Computer systems interpret data in real time. Data may also be stored and subjected to further analysis at a later time; as a result, a modern turbine undergoes continuous testing during its normal operations.

Below is a list of common turbine problems.

- Cracking in turbine case: Cracks are caused by thermal stress.
- Blade chipping: Chipping is caused by particles emitted from the boiler. The inner surface of boiler pipes can oxidize, creating a rustlike coating. The

coating sheds particles that impinge upon the turbine blades and cause erosion.

- Blade failures: Resonant vibrations cause mechanical stress in blades near inlet.
- Sealed bearing failure: Vibrations during start-up cause stress on the sealed bearings that seal the shaft as it detrudes from the turbine. This creates leaks in the seal.

2.3.2 Coal Steam Generation, Operations, and Environmental Considerations

This section discusses issues that are particular to coal-fired generation. These issues include operations, emissions and environmental concerns, and hazardous emission mitigation technologies.

Operations

Coal is the most abundant and economical fuel source available in the US. Around 50 percent of electricity output in the US is fueled by coal. Coal-exporting states include Tennessee, Kentucky, West Virginia, Virginia, Pennsylvania, and Wyoming. Coal is transported from mining states to consuming states via rail. Many utilities have long-term relationships with specific mines and own the rail systems between the mines and the generation plants. There are several coal technologies. A technology known as pulverized coal combustion is the most common and is described below.

At each coal plant there is normally an inventory of coal. Typical inventories can supply a station for 2 months. The inventories are nothing more than piles that are stored in open space. Each day, coal ready for burning is placed in a hopper. The hopper dumps coal onto a conveyer belt. The conveyer belt then loads a mill, which crushes the coal into fine dust. The coal dust particles are extremely tiny; 75 percent have diameters less than 75 micrometers, much smaller than a dot made with a sharp pencil. Milling the coal into fine dust increases its surface area, which enhances combustion efficiency. Coal dust exits the mill where fans blow the dust into the furnace. There it is burned at high temperatures, 2800 degrees F.

Natural and induced draft force the by-products of the burned coal past the boiler and up the chimney stack. Unfortunately, there are many toxic and hazardous elements, both gaseous and solid, among these by-products.

Toxins and Hazardous Emissions

Coal is among the most environmentally unfriendly fuel sources. Environmental degradation occurs because of mining activities; additionally, hazardous emissions result from the burning of coal. Below is a list of hazardous emissions.

- Nitrogen oxides (NOx): NOx contributes to ground-level ozone and smog through chemical reactions with other airborne elements. Excessive ozone is unhealthy for human and animal tissue. NOx is also a greenhouse gas.
- Sulfur dioxide (SO_2): SO_2 emissions are responsible for the phenomenon of acid rain. High acidity in soil and lakes is a known hazard for animal and plant species. Additionally, SO_2 adversely affects human health. Although the scientific community agrees on the adverse effects of SO_2 emissions, quantifying their impact has been difficult and there is no consensus on the degree of harm to ecosystems and human health. SO_2 is emitted as a gas.
- Fine particles ($PM_{2.5}$): Coal emissions include fine particulate matter that is harmful to human health. This includes sulfates as well as particles composed of organic compounds (carbon-based compounds). Carbon-based compounds are of greatest concern to human health because of their known carcinogenic properties.
- Mercury: Mercury is a toxic metal that is found in coal. The level of mercury in coal that is emitted from smokestacks as vapor is a concern. Mercury is inert; once it has established itself in the ecosystem it remains indefinitely. Continued burning of coal increases the pool of mercury in the ecosystem. Because of the density of mercury, it precipitates out of the atmosphere readily and does not disperse. This creates localized areas of contamination levels that are of concern. Mercury is known to damage neural, kidney, and other organ functions in high concentrations such as exposure to industrial processes with improper ventilation.
- Greenhouse gas (CO_2): Carbon dioxide causes what is known as the greenhouse gas effect. The Earth receives and emits radiant energy. The portion of the Earth's surface exposed to the sun during the daytime receives more radiant energy than it emits, causing surface heating. Conversely, during nighttime hours the surface emits more radiant energy than it receives, causing surface cooling. CO_2 traps a portion of the radiant energy that the Earth emits. Scientists believe that an increase in the amount of CO_2 in the atmosphere will shift the radiant energy balance and cause global warming. Coal combustion is one of the largest contributors to increased atmospheric CO_2.

Emissions Reduction Technologies

Several technologies are available for reducing toxic emissions. Different technologies address the emissions of different toxins and hazardous elements. Two common technologies are listed below.

- Selective catalytic reduction (SCR): Selective catalytic reduction is used to reduce NOx emissions and works much like a catalytic converter on a car. Flue gas containing the NOx emissions from the combustion process is mixed with ammonia. The mixed gases travel through a series of catalytic layers, which cause the NOx to react with the ammonia. The reaction converts the

NOx to pure nitrogen, a benign chemical that makes up 80 percent of the air we breathe, and water vapor. Both elements are returned to the environment through the station's stacks.

- Wet scrubbers: Wet scrubbers remove up to 95 percent of the SO_2 emissions as well as achieving reductions in emissions of particulate matter. Wet scrubbers operate by spraying a mixture of pulverized limestone and water into the exhaust gas of the generating units. Inside the scrubber vessels, calcium in the limestone reacts with the gaseous SO_2 to form calcium sulfate, commonly known as gypsum, and water vapor. In addition to SO_2, carbon-based particulate matter is removed from the flue. The gypsum and particulate matter are collected and stored.

Remarks

- The features of coal plants described in this section have maintenance and operational issues in addition to those of generic steam turbine generation. Every component and process along the fuel path, from mine to chimney exit, has special operational characteristics and is vulnerable to breakdown. Two interesting examples of complications occur because of frozen coal.
    - Coal can freeze in the railcar on its way from the mines to the power plant. This creates difficulties in unloading the cars. The normal unloading procedure is that the car has a rolling mechanism that allows the car to tip sideways and dump the coal out. When the coal is frozen, this is not possible.
    - Frozen coal in the inventory creates difficulties because it cannot be loaded into the hopper. To prevent freezing, antifreeze similar to the chemical used in cars is sprayed over the coal inventory. In case of freezing, dynamiting is used to loosen the pile.

2.3.3 Nuclear Steam Generation

This section presents the basic elements of nuclear steam generation. The section begins with a description of the reaction process and then outlines the process of converting reaction energy to steam. Refueling of nuclear reactors is a process that affects the operations and economics of the plant; accordingly, the refueling cycle and procedure are presented. Safety precautions peculiar to nuclear generation are also presented.

Nuclear Energy

There are two categories of nuclear energy, fusion and fission. Fusion is the process of converting atoms of one type into another by fusing their nuclei. The energy state of the fused atoms is lower than the energy state of the original atoms. A common fusion process in the sun is the fusion of two hydrogen atoms to form a helium atom. The differential in energy states is released as radiation. It is hoped that this energy can be harnessed for power production. However, the engineering difficulties associated with fusion have not been overcome and there are no fusion generators.

The other category of nuclear energy is fission. Fission occurs when an atom in an unstable configuration splits. On splitting, new atoms are produced. As with fusion the energy state of the newly created atoms is lower than the energy state of the original atom. This energy difference is available for power production.

The specific fuel for use in standard nuclear fission reactors is uranium. Uranium is a heavy atom with 92 protons in its nucleus. Uranium is present in two forms, U-238 and U-235. U-238 and U-235 differ in the number of neutrons they possess: 146 within a U-238 atom and 143 within a U-235 atom. The extra neutrons provide stability in the U-238 nucleus. On the other hand, U-235 is unstable and subject to decay through fission. Natural deposits of uranium contain 99.3 percent U-238 and only 0.7 percent U-235. Since U-235 is subject to fission, this is the material that fuels nuclear reactors.

The concentration of U-235 in unprocessed uranium is too low to be suitable as a fuel source. Useful fuel must have U-235 concentration levels above 4 percent. (Weapons-grade uranium requires U-235 concentrations of 95 percent.) There are several processes available for increasing U-235 concentration. A common method uses centrifuges to segregate the lighter U-235 atoms from the U-238 atoms. The process is called enrichment.

After enrichment, the fuel is packed into pellets and placed into the fuel rods of a power reactor. Fission of a single U-235 atom releases a small amount of energy. Newly formed atoms with atomic weights below U-235 inherit 85 percent of this energy in the form of kinetic energy. Subatomic particles carry the remaining 15 percent. A natural chain reaction promotes continued fission. Neutrons released from one U-235 atom bombard neighboring U-235 atoms, causing subsequent fissioning of neighboring atoms. Sustained and stable reactions require that on average each fission of a U-235 atom causes one subsequent fission in neighboring atoms. Energy release would diminish if each fission caused less than one subsequent fission and would increase to dangerous levels if each fission caused more than one subsequent fission. Sustainable reactions are not possible if the concentration of U-235 atoms is insufficient because the neutrons released from fission would have too low a probability of striking another U-235 atom to cause a subsequent fission. This is why fuel enrichment is necessary.

As the fission rate is critical to controlling power output and safety, reactors have a special mechanism to control this rate. The mechanism relies on a system of control rods that are able to absorb neutrons to slow down reaction rates. An engineer may lower the control rods between the fuel rods to slow down reaction rates. Conversely, lifting the control rods allows neutrons to bombard U-235 atoms in neighboring fuel rods; this increases the reaction rate. By adjusting the rods to the appropriate level, a stable reaction rate can be achieved and maintained.

Control rods are made of cadmium and boron. Each of these substances has neutron-absorbing properties. In the event of an emergency, all control rods can be quickly lowered into the core. This slows the reaction by lowering the reaction rate

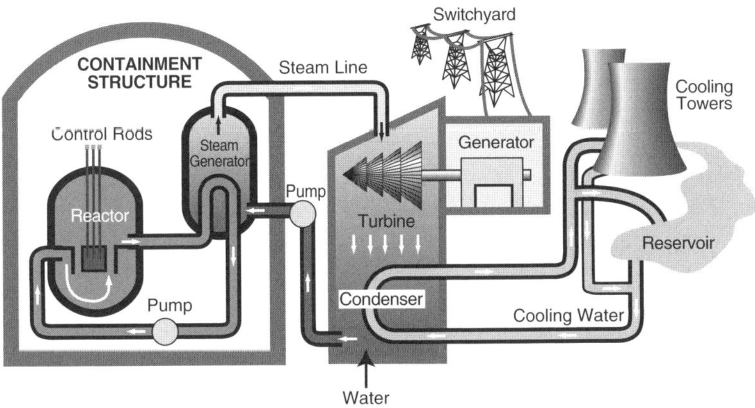

Figure 2.6

well below the sustainable level. Slowing the reaction to a desired level may take a considerable amount of time, as evidenced by the Three Mile Island incident. This incident is addressed below.

A central concern of nuclear plant operators is the temperature within the reactor core. Without dispersion of the reaction energy, heating primarily caused by the kinetic energy of fission products would cause fuel rods to melt and radioactive material would be released outside the core. There are several system designs for maintaining core temperatures at safe levels. The pressurized water reactor is most commonly used in the US. The remainder of this section discusses this system. Figure 2.6 illustrates a pressurized water reactor.

The pressurized water reactor contains a primary and a secondary circulatory loop. Water within the primary loop enters the core, where it is heated up. From there it is circulated to a heat exchanger that provides the heat source for steam generation in the secondary loop. Water in the primary loop only comes in contact with the reactor, not the turbine or the condenser. Water and steam in the secondary loop only come into contact with the turbine and the condenser, not the reactor core. The heat exchanger takes on the role of a boiler in a coal plant. This is where steam that drives the turbine is heated up.

Under normal operating conditions the water of the primary loop cools the core sufficiently for maintaining safe core temperatures. Since maintaining the core temperature is of paramount concern, there is a redundant cooling system consisting of a water reservoir that can be pumped into the primary loop.

Refueling

Refueling at a nuclear plant typically occurs every 18 to 24 months. Only 20 years ago refueling required 2 to 3 months, during which the plant was inoperable. Currently, engineers can refuel a plant in 3 to 4 weeks. Refueling outages may be delayed in order to ensure that a plant is available during seasonal demand peaks. For example, one may reduce the output of a plant during the winter and spring to delay a summertime refueling outage.

Nuclear fuel is packed into the core in units called assemblies. As power is produced, the concentration of U-235 in each assembly is reduced. Refueling requires replacement of the depleted assemblies, those in which the concentration of U-235 is too low for sustained reaction. Assemblies are packed into the core in cycles. One-fourth of the assemblies are replaced each refueling period.

Aside from replacing depleted assemblies, other assemblies are reshuffled to different cell locations during refueling. Reshuffling of assemblies is done to ensure even heating throughout the core. This requires that the rods with similar U-235 concentrations are scattered throughout the core.

Disposal and Environmental Concerns

The spent fuel is stored at an on-site facility. The material is radioactive and includes plutonium. Fission continues to occur at a rate that generates heat; the rods are therefore placed in a water reservoir. Currently there is no permanent solution for managing spent fuel in the US. It is hoped that a permanent storage location will become available for receiving radioactive waste. Indeed, Yucca Mountain in Nevada has been identified as a desirable location because it provides a geologically stable environment. Because of political opposition, this has not yet come about.

The main environmental concern associated with nuclear power generation is the management of radioactive waste. It should be noted that nuclear power produces no toxic emissions or greenhouse gasses. The balance of risk between nuclear waste management, toxic emissions, and greenhouse gas emissions is central to the policy debate concerning the future of nuclear energy.

Remark

- There are other nuclear reactor designs aside from that presented above. One well-known reactor is the breeder reactor. Within the breeder reactor there is a cache of U-238 that envelopes the reactor core. Neutrons from the fissioning U-235 penetrate the U-238 nucleus, converting it to U-239. The U-239 emits two beta particles, converting two neutrons into protons. The final product is plutonium, a fissionable material that can be used as nuclear fuel.

Three Mile Island

The worst nuclear accident in the US occurred in March of 1979 at the Three Mile Island plant in Pennsylvania. Since this accident many nuclear projects have been abandoned, and there have been no applications to build more plants. Below is a brief chronology of the accident. The account indicates standard safety features as well as their failure at Three Mile Island. The account is based on the chronology available at http://www.pbs.org/.

Before providing the account, one bit of information is necessary. Nuclear reaction slows down, but does not cease, with the lowering of control rods. Although the reaction is no longer indefinitely sustainable, fission continues at a rate that produces significant heat for several days.

1. The First Minute A malfunction causes an automatic shutdown of the pumps that circulate fluid through the secondary loop. Fluid from the secondary loop no longer cools the fluid in the primary loop, and temperature as well as pressure within the primary loop rise. A pressure relief valve within the primary loop opens to release steam. Unknown to plant operators, the valve malfunctions and does not close, allowing steam within the primary loop to exit through the valve. The operators also are unaware that another set of valves that guide the fluid in the secondary loop toward a backup set of pumps are closed. Operators erroneously believe that fluid in the secondary loop is once again circulating and cooling down the fluid in the primary loop. The control rods automatically lower in response to the temperature increase within the primary loop.

2. Minutes 2 Through 8 A valve allowing water to enter the primary loop from the redundant reservoir opens, and water enters the primary loop. Operators respond to a rise in the water level and decrease in pressure within the primary loop by closing the valve to the redundant reservoir. They are unaware that the pressure decrease is due to steam exiting from the malfunctioning pressure relief valve. With steam exiting from the primary loop, the water level decreases and the core temperature continues to rise. One operator notices that the valves that guide fluid to the secondary loop's backup pumps are closed and opens these valves. Fluid is now circulating through the secondary loop.

3. Minutes 10 Through 45 Monitoring systems fail to indicate that steam is exiting from the pressure relief valve. An alarm that should have been triggered by rising levels of radioactivity associated with the release of steam from the primary loop does not go off. Gauges that monitor the water level within the primary loop provide conflicting readings. Operators interpret the readings as indicating that water levels are not low.

4. Hours 1 Through 4 The pumping system within the primary loop is disabled. Water no longer circulates through the primary loop, resulting in accelerated heating of the core. Eventually as water converts to steam, the core is no longer covered by water. Hydrogen as well as radioactive gases exit through the malfunctioning pressure relief valve. A radiation alarm sounds, and authorities declare a site emergency. This is soon elevated to a general emergency with indications of increased levels of radioactivity.

5. Hours 7 Through 16 A shift change brings about a new arrival; the individual assesses that the relief valve is malfunctioning. Using redundant systems, the relief valve is bypassed and steam is no longer exiting from the primary loop. The valves allowing water to enter the primary loop from the redundant reservoir are once again opened. Water enters the primary loop; however, by this time the core temperature is excessive and the water vaporizes. This causes dangerously high pressure within the primary loop, so a second relief valve is opened and the valves to the redundant reservoir are closed. Once again, steam exits out from the relief valve along with hydrogen gas and radioactive material. After a long wait, it is judged that the redundant reservoir valves can once more be opened. Water once again enters the primary loop, and the core begins to cool.

6. Hours 16 Through 72 There is concern that an explosion of a hydrogen bubble in the containment vessel would rupture the vessel and radioactive material would spew from the rupture. To avert this scenario officials vent gas from the containment vessel, releasing radioactive material. Although radioactive gas is released, the far more dangerous scenario is averted.

7. Aftermath A cleanup effort costing around $973 million began in August 1979 and ended in December 1993. Since the accident, within the US no new orders for nuclear reactors have been placed with manufacturers and many were canceled. Only one new reactor that was under construction at the time of the accident has come online. Although economic factors were the central cause for discontinuing investments in nuclear facitlities, the accident hastened decisions and created a political environment that looked on nuclear energy with disfavor.

2.3.4 Combustion Turbines

This section takes a detour from steam generation technologies to discuss combustion turbines (CTs). The detour is necessary before discussing combined cycle gas turbines (CCGTs), another steam generation technology, because output of hot exhaust gases from CTs heats the steam of a CCGT's steam generator. This section includes a description of CT components and their functionality and efficiency, operations, maintenance, and emissions.

How They Work

There are similarities between combustion turbines and steam turbines. In a steam turbine, steam moving from high pressure to low pressure causes the turbine to rotate. In a combustion turbine, air moving from high pressure to low pressure causes the turbine to rotate. Components of the combustion turbine are presented in Figure 2.7. Air enters the turbine through an inlet valve. There it is compressed and attains a high pressure, 30 atmospheres. From the compressor air moves to the combustor, where it mixes with fuel and ignites. Temperatures range between 1200 and 3000 degrees F, depending on equipment and operating levels. The ignition causes an expansion in volume, and the air moves to lower pressure, passing through the turbine. The turbine rotates and spins the electric generator (indicated as power

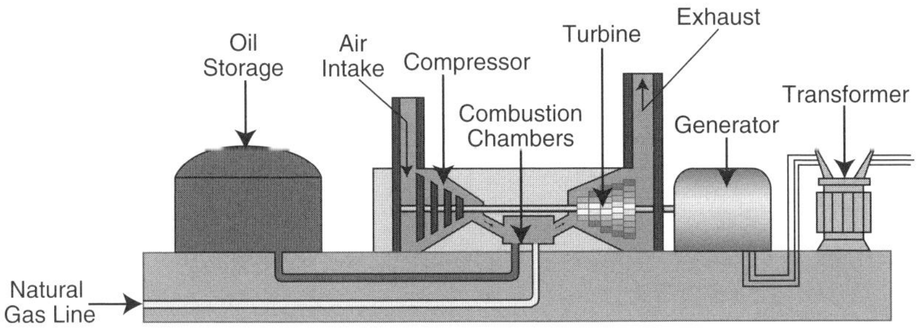

Figure 2.7

turbine) as well as the compressor. Indeed, 67 percent of the power produced by the turbine is used to power the compressor, while the remaining 33 percent produces electric power.

Remarks

- Combustion turbines as described above have optimal heat rates between 10.2 and 12.5 MMBTU/MWH.
- Efficiency may be enhanced by several techniques. One technique channels the exhaust gas back to the combustion chamber, where it is preheated. Another technology cools air in the compressor to increase the efficiency of the compressor.
- With such enhancements, the most efficient combustion turbines have heat rates at 9.3 MMBTU/MWH.

Operations

Natural gas sources for the US are primarily located in Texas, the Gulf of Mexico, and western Canada. Additional deposits are present in New Mexico, Colorado, and Wyoming. A network of gas transmission pipes transports natural gas from the wellhead to the consumer. Aside from piped natural gas, a small amount of liquefied natural gas is imported to the US from overseas sources.

Any gas plant must couple to the gas transmission network to ensure supply. Gas flow rates and pressure within the pipes are regulated by electronic pumping mechanisms. Once at the power plant, pumps near the combustion chamber regulate flow rates in the combustion chamber. This regulation controls the combustion rate and power output.

The operations of a CT differ from those of a steam turbine. Start-up times are much shorter; A CT may be online within 10 minutes of commencing operations. For this reason, CTs are ideal as resources for reserves. Starting up a CT requires an auxiliary motor that is not shown in the diagram or picture. The auxiliary motor operates the condenser before the turbine is able to assume that function. (Before a turbine can operate there must be sufficient pressure in the compressor;

otherwise, air would blow out through the intake on ignition and a unidirectional airflow would not be established.) The auxiliary motor also initiates rotation of the turbine.

Ignition in the combustor commences after sufficient pressure is attained in the condenser. Power output is regulated by controlling the combustion rate within the combustion chamber. CTs are often operated remotely through automated systems.

Emissions

The fuel for CTs is natural gas. Natural gas is a favored fuel because there are no particulate emissions or trace metals. Hazardous emissions are restricted to NOx and CO_2, and the emissions of these products from natural gas are much lower than for coal. These attributes make natural gas a very attractive fuel source, and most new generation built since 1995 is fueled by natural gas.

Some CTs have dual fuel capability, heating oil can be a secondary fuel. From an emissions standpoint, heating oil is a less attractive fuel than natural gas; burning heating oil produces SO_2 as well as increased NOx emissions.

Turbine Blade Replacement

A CT has two to four rows of turbine blades. Each row has 75 to 150 blades. Turbine blades fail because of mechanical and thermal stress. CT blades are much longer than steam turbine blades. For larger turbines, the tip of the blade is up to 5 feet from the center of rotation. The blade itself rests on a cylindrical shaft and protrudes up to 30 inches from the cylinder. The velocity at the tip of the blade is very high, reaching speeds in excess of 1200 miles per hour. Not surprisingly, there is greater mechanical stress on CT blades than on steam turbine blades. Thermal stress is also of greater concern in CTs than steam turbines, primarily because the temperature environment is more extreme. Temperatures of a combustion turbine reach up to 3000 degrees F vs. 1500 degrees F in a steam generation turbine. Because of the stress on a CT, turbines require periodic total blade replacement as prescribed by the manufacturer. Both starting up and shutting off a CT cause considerable turbine stress.

Besides start-ups and shutdowns, trips (mechanical failures that cause a unit to quickly cease functioning) cause excessive stress. Blade replacement protocols are based on the number of start-ups, the number of trips, and the number of operating hours. The manufacturer relates unit start-ups and unit trips to an equivalent number of operating hours. Replacements are set on the basis of equivalent running hours. For a frequently used unit, replacements may be required every 5 years. A CT that drives a CCGT, as described in the next section, is frequently used.

2.3.5 Combined Cycle Gas Turbines and Cogeneration

The exhaust of a CT contains much waste energy. This energy can be used for the generation of electricity as well as other industrial processes. The use of exhaust energy for industrial purposes is called cogeneration. This section describes facilities that generate power from the energy in the exhaust of a CT and provides an example of cogeneration.

Combined Cycle Gas Turbine

A combined cycle gas turbine (CCGT) utilizes high-temperature exhaust from one or two CTs to produce additional electric power. After fueling a CT, the exhaust gas exits the CT at 1500 degrees F and enters a heat recovery steam generator (HRSG) (Fig. 2.8). The HRSG acts a boiler; water is converted to steam within the HRSG, using the heat from the CT's exhaust gas. The steam powers a steam turbine as described in Section 2.3.1.

Efficiency gains from operating in CCGT mode are substantial. A typical stand-alone turbine has an optimal heat rate of 10.5 MMBTU/MWH (efficiency of 32%). In CCGT mode, the heat rate can be below 7.0 MMBTU/MWH. Indeed, there are claims of CCGTs with heat rates of 6.5 MMBTU/MWH (efficiency 53%). Fuel consumption of a CCGT is 62 percent of that of a CT for the same energy output.

Remarks

- In previous discussions of emissions, efficiency was not mentioned. But the impact on emission levels and emission reduction is straightforward. Less fuel burn translates directly into lower emissions.
- Emissions reduction technologies are at times used with CCGTs to reduce NOx emissions.

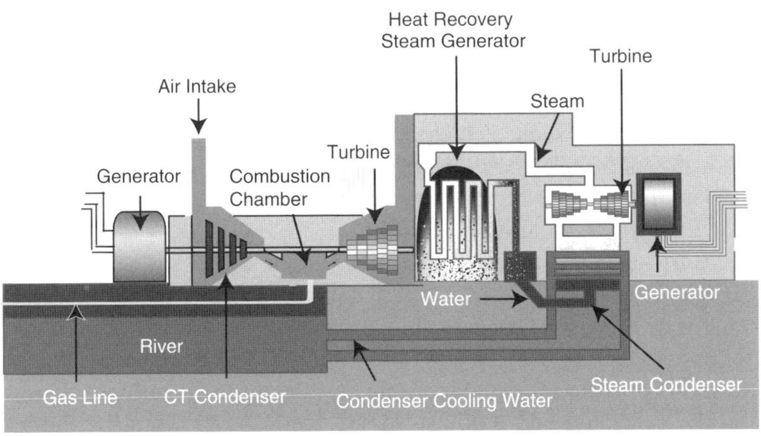

Figure 2.8

Cogeneration

There are many industrial processes that require steam. It is often cost-effective to utilize a CT to generate steam for industrial processes and generate electricity at the same time. The configuration is similar to that of a CCGT. There is a HRSG that utilizes the waste heat of the CT for generating steam. However, instead of this steam being used to power another electric generator, it is used for an industrial process.

An example occurs in oil extraction. To increase extraction rates steam is pumped into an oil field. The steam induces pressure that increases the flow rate of the oil. The waste heat of CTs may be used to produce the steam.

2.3.6 Hydroelectric Power

Hydroelectric systems provide another source of power generation. In the US, hydroelectric systems account for around 9 percent of electric energy production. This section describes the hydroelectric generation process along with operational, maintenance, and environmental issues. Three types of hydro facilities are discussed: large dams, run of river, and mountain networks. Some details about Hoover Dam are provided to illustrate concepts.

Generation Process

Major hydroelectric systems are located in Washington state, Nevada, California, and Tennessee. These regions provide abundant water flows in mountainous settings. Hydro systems convert the kinetic energy of flowing water into electricity by channeling the water through a turbine, causing the turbine to rotate. The turbine then spins an electric generator. The available power is a function of the elevation differential between the surface of the water source and the turbine; larger elevation differential provides more power.

To capture energy associated with elevation differentials, the largest hydroelectric systems build dams. The dams create a lake behind them. The surface elevation of the lake is high in comparison with the bottom of the dam. A system with elevation h and flow rate R pounds of water per second has a potential power output of $h \times R$ ft-lb/s The surface water of Lake Meade behind Hoover Dam is on average 520 feet above its outlet. The flow rate of water through its systems varies but can be up to 3.54 million pounds of water per second. The potential for power production is 1.84 billion ft-lb/s or around 2500 MW.

Hydroelectric energy conversion is far more efficient than any other production method. Hydro systems can convert 80 percent of the available energy into electricity. This efficiency is possible because, unlike fossil fuel generation, there is no heat loss out the smokestack or exhaust loss. Also, there is no loss in the conversion of water to steam. Losses are restricted to mechanical losses in the turbine, microwave losses from the generator, and frictional losses due to water viscosity. Figure 2.9 provides a diagram of a large-scale dam project.

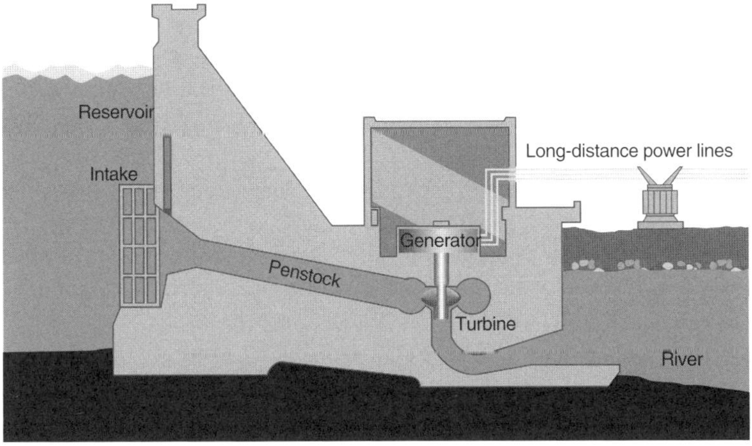

Figure 2.9

Hoover Dam: Configuration and Operations

At Hoover Dam there are 19 turbines, each with its own generator. The turbines are housed in buildings on the banks of the river below the dam. Ten units are on one side of the river, and nine units are on the other side.

There are four major penstocks leading from the reservoir to the river, two on each side of the riverbanks. Each penstock has its separate entryway. The entryways are visible on the lake's surface as towers. The towers have gates to seal the entryway to the penstocks. These gates are closed only during maintenance and inspection of equipment between the penstocks and outlet: turbines and valves. At all other times the tower gates are open.

For each turbine there is a channel that leads from the penstock to the turbine. Water flows down these channels to spin the turbines. Within the channel close to the turbine is a valve called a butterfly valve. The butterfly valve is used to regulate the flow rate from the reservoir. Closing all 19 butterfly valves completely interrupts flow from the reservoir to the river below.

After passing through the butterfly valve, water enters the turbine tunnel. The tunnel winds and narrows as it circles around the turbine. Surrounding the turbine blades are wicket gates, which regulate power output. The wicket gates guide the water toward the turbine blades to provide more turbine power or away from the turbine blades to provide less turbine power. The turbine itself spins around a vertical axis. The electric generator is mounted on this vertical axis and spins with the turbine. After passing the turbine water flows vertically downward, passes through a bend, and then exits the passageway to the river.

One is able to alter the output of hydroelectricity faster than any other type of generation technology by controlling the wicket gates. For this reason, operators use hydro generation to respond to supply and load conditions on the grid.

Remarks

- Hydro generation often competes with many other social interests for the rights to determine water flow levels. Other interests include agriculture, municipalities downstream from the dam, recreational requirements, and environmental impact. The legal rights to flow activities do not always match social needs.
- Weather affects holders of legal rights. These holders frequently respond to weather conditions without regard to hydro generation benefits and needs. The Hoover Dam operations and associated water rights comprise one of the more complicated sets of water usage arrangements made in times when men "drank whiskey and fought over water rights." There has been little change.
- Anyone hiking the Grand Canyon welcomes the sight of beautiful turquoise water at the canyon's bottom. The first instinct is to dive, but one recoils instantly after a toe check. Water flowing through the Grand Canyon passes through Hoover Dam. Hoover Dam releases water from the bottom of its reservoir. This water receives no sunlight and is very chilly; the temperature is around 43 degrees F. The cold water has changed the ecosystem within the Colorado River downstream from Hoover Dam. Additionally, this water is nearly silt free, as silt does not flow through the reservoir. The silt-free river has caused erosion along the riverbanks downstream from Hoover Dam. Environmentalists raise these concerns in public forums.
- The main competing interests on the Columbia River in Washington are those of the fishing industry and hydro generation. Salmon migrate up the Columbia to spawn. Hydroelectric facilities are tremendous obstacles for salmon to overcome and affect the their ability to make it to their spawning destination. Cheap electric power as provided by facilities along the Columbia River is critical to many industrial interests and desired by many residents. There have been efforts to accommodate fishing interests, such as building salmon ladders and elevators. Nevertheless, tensions between these different stakeholders for water system usage continue.

Run of River

Another type of hydro system is a run of river system. In this system a barrier is placed across a river to create a height differential between the upstream inlet and the downstream outlet. However, the system does not create an expansive reservoir.

Because there is no reservoir behind a run of river facility, it is more difficult to manage flow rates through the dam. Indeed, power output ebbs with the natural river current.

Mountain River Systems and Reservoir Management

Another favorable location for a hydro system is along mountain streams. Many mountain stream hydro systems are characterized by a series of cascading lakes that form reservoirs for the hydro system. The lakes can be natural, enlarged with dams, or created by dams. Each lake has an associated penstock that runs down the mountainside and leads to a single or multiple turbines.

Mountain systems rely on the springtime melting of the winter snow pack to replenish water levels annually. Accordingly, the biggest impact on reservoir levels is the winter snowfall. The second biggest impact is the rate of springtime thawing. A short snowmelt season adversely affects mountain networks because reservoirs are overwhelmed and engineers must open spill gates. Potential energy from the spilled water is lost, and the major source of reservoir replenishment ceases early in the year, meaning that the hydro resource will not be able to provide power as planned. Hydro generation conditions are most favorable when there is abundant snowfall with a long, slow snowmelt.

Hydro systems generally benefit from rainfall as rainfall provides supplementary water replenishment. However, rainfall is not always beneficial. Indeed, a heavy rainfall that causes significant snowmelt may adversely impact a hydro system as described in the preceding paragraph.

Pumped Storage

Recall the load shapes discussed in Section 2.2. There is great load fluctuation. Imagine if storage facilities were available. Maintenance problems would be reduced because there would be less stress induced by starting and stopping equipment. Whenever supply output was higher than load demand, the surplus would be stored in a storage facility. Whenever the load demand was higher than the supply output, the deficit would be released from the storage facility. Under such a scenario fewer plants would have to be built and the plants providing power would all be very efficient.

There is only one practical storage method available, pumped storage (Fig. 2.10). Unfortunately, this has limited use because of geographic limitations. Never-

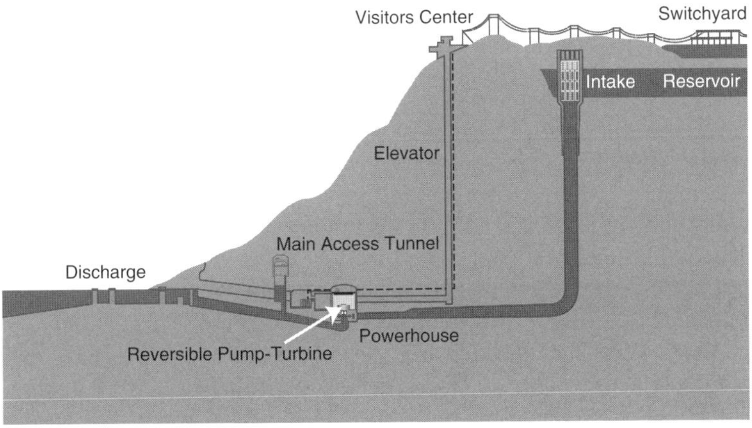

Figure 2.10

theless, wherever it is practical, pumped storage is very valuable. A pumped storage facility is a single mountainside hydro system in which there is limited or no inflow into the reservoir. At nighttime, when loads are low, the generator acts as a motor and the turbine acts as a pump. Water is pumped up the mountainside back to the reservoir. During the daytime when loads are high, the water is released from the reservoir, providing hydroelectric power.

Maintenance and Performance

Hydro systems require less maintenance and have fewer forced outages than nuclear and fossil fuel systems. Several factors account for this. First, there is less equipment so less can go wrong. There are no boilers, heat exchangers, compressors, fuel mills, fuel pumps, control rods, condensers, and other system elements associated with other generation plants. Additionally, the operating environment of hydro facilities is much less taxing on equipment than the operating environment for nuclear and fossil fuel systems. There is no thermal stress on the turbine and other components. Lifetimes are accordingly longer, maintenance requirements are less, and forced outages are fewer.

Hydro systems require regular maintenance of equipment. This includes lubrication of valves, turbine maintenance, and generation maintenance. Inspections to ensure the integrity of the dam structure and penstocks are also necessary.

The major uncertainty in hydro system performance is the weather. With the exception of pumped storage systems, engineers have no influence on the water that is available to the system. As mentioned above, seasonal as well as daily weather uncertainties can cause uncertainty in system performance. For all hydro systems, changing climatic conditions can have significant impact and there is much uncertainty. Drought conditions in the West have caused shrinkage in the size of Lake Meade behind Hoover Dam. The surface level is currently 200 feet lower than it was at its peak. Accordingly, less power is available. Even more sobering is the forecast that the surface level will fall another 300 feet by 2017; from that point on, Hoover Dam will not be able to generate power in its current configuration. When planners first designed the Hoover Dam generation system, they did not believe that such a significant drop in the water level would occur.

2.3.7 Wind Power

A visit to the Dutch countryside provides the opportunity to see the impact of harnessing wind energy from a historical perspective. The Dutch built a country out of wind in a literal sense. Much of the country lies below sea level; what is now land was once swamp and lakes. Since the 1500s the Dutch have established communities of farmers who have cleared the lakes and swamps to develop farmlands. The Dutch etched canals throughout the country. They segregated the canals from the farmland, using mounds that act as barrier walls by containing the canal water. Within the farmland are channels leading to the canals. The Dutch placed windmills atop the

mounds and used the windmills to power water wheels that pump water up through the channels and into the canals. The Dutch have preserved some historical sites such as the one at Kinderdijk pictured above.

This system is still very much in operation: however, it has been modernized. Electric pumps have taken the place of waterwheels. This is not the end of wind power for the Dutch. There is a national effort to harness wind power for electric generation. So a growing portion of the energy used to power the electric pumps is in fact wind energy.

In the US wind generation accounts for 0.15 percent of power production. Some states such as California have an ongoing effort to promote wind generation. In southern California wind accounts for 1.5 percent of energy production. This section reviews wind generation technology along with operational issues.

Wind Turbines

Wind generation, like all conventional generation technologies, uses a turbine to rotate a generator. And as with all conventional technologies, a fluid moving from high to low pressure (the wind) rotates a turbine.

The components and functions of a wind turbine are given below (see Fig. 2.11).

- Internal computer systems: Internal computer systems address three functions. First, they perform diagnostics on other windmill components and correct or shut down any system problems that they identify. Second, they communicate with external operators, providing system status as well as executing commands from the external operators. Third, they form a complete monitoring and control system for optimal blade operations. They monitor wind speed and direction, calculate optimal blade configuration, and operate systems to align blade operations with the optimal configuration.

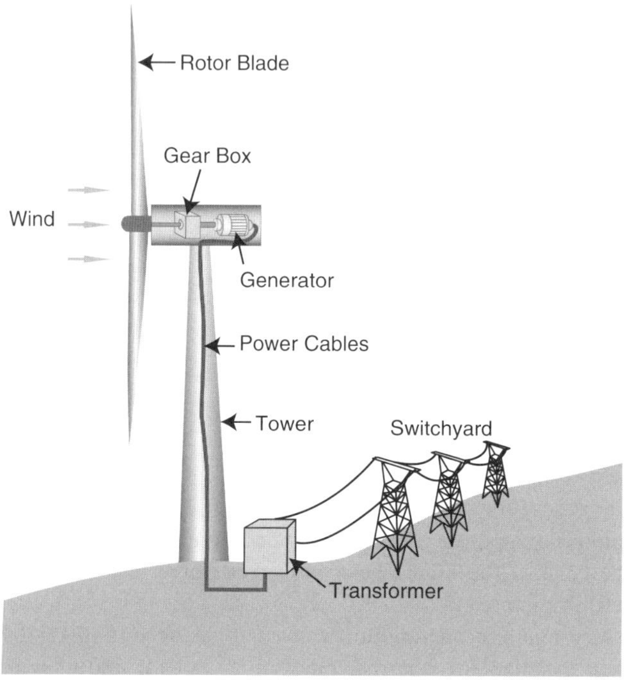

Figure 2.11

- Internal motors: An internal pointing motor, the yaw motor, receives commands from the computer system and points the blades in the direction of the incoming wind. Another motor adjusts the pitch of the blades in accordance with commands from the computer system.
- Turbine: The turbine consists of blades and a locking mechanism. The pitch of the blades is adjustable to adapt to different wind conditions. At high wind speeds the blades are tilted to reduce their angle into the wind. A locking mechanism prevents the blades from rotating during periods of excessive wind speed, above 55 miles per hour. Excessive rotational speeds overheat the generator.
- Gearing mechanism: A gearing mechanism couples the turbine with the generator. The turbine rotates at low speeds, 11 to 20 rotations per minute. The turbine shaft is connected to the gearing mechanism, through which the rotational speed is increased to 1200 rotations per minute as required by the generator.
- Electric generator: Wind generators are currently much smaller than those coupled to other facilities. A 3-MW wind generator is considered very large.

- Output control systems: Wind generation output is very unsteady and must be stabilized to meet grid requirements. Grid facilities must be installed to safely integrate wind generation output into transmission lines. Such facilities are expensive.

Despite centuries of harnessing wind energy, the use of wind for generation in commercial quantities as a supply source is very recent. The technology is still maturing. Technical problems with producing large-scale output (500 kW and greater) arise for two main reasons: design requirements of the turbine and network fluctuations due to variable wind.

Unlike steam and combustion turbines, wind turbine blades must be lightweight to perform well. Wind turbine blades are subject to oscillations from fluctuating winds that create significant mechanical stress. So in addition to being lightweight, turbine blades must be strong and must have systems that reduce the stress of oscillations.

There have been continued research and development efforts to improve turbine design. Although the mechanical problems appear to be solved, output fluctuations affecting grid reliability currently place a limit on the percentage of energy output from wind at around 20 percent. Efforts to provide more wind power would cause instability in transmission lines that has the potential to create blackouts. Output fluctuations as evidenced by a variance in voltage and power are caused by unsteady wind conditions.

Regulated utilities typically must demonstrate sufficient resources to meet projected demand. This is accomplished by summing the total generation capacity and comparing it with the forecasted load. Because of the unsteady and unpredictable nature of the wind, there is much debate concerning the accounting of wind power in such calculations. Typically, the projection of output is drastically discounted in order to account for days that the wind does not blow. The effect is that additional reserve capacity must be available during for the days that wind turbines are not able to generate power.

2.3.8 Summary of Technologies

This section provides cost information for construction of the technologies addressed above. It then grades each technology in several categories: cost of construction, cost of operations, reliability, flexibility, availability of fuel source, emissions, and environmental impact. From the results it is apparent that there are many complicated trade-offs that must be addressed in determining the correct resource base for future power generation. These trade-offs can only be addressed through policy that comes about through open public debate.

Before proceeding we would like to mention that there are other technologies that currently don't play a significant role in energy production but might in the future. These include several coal alternatives, different solar technologies, and fuel cells.

Cost of Construction

The industry standard for communicating costs is in \$/kW of capacity; we adopt this standard. Table 2.5 presents costs for different technologies and serves two purposes. First, it gives a rough idea of costs of generic projects, and second, it provides a basis for comparing costs between different technologies. Table 2.5 is more successful with the second objective. The cost of construction for projects varies greatly; a prime reason for differences is the cost of land.

Characteristics

Table 2.6 provides output and optimal heat rate characteristics for the different technologies.

Table 2.7 grades the technologies on various characteristics. The grades range from A to E, with an A grade indicating the most socially favorable rating for a given characteristic. These ratings are of course subjective and not meant to attribute a definitive grade for the characteristics. Table 2.7 has greater value as a source of debate. Readers should challenge the grades, providing reasons for either agreeing or disagreeing with the grade.

Table 2.5 Cost of Construction

Technology	Cost of construction \$/kW
Pulverized coal	1600
CCGT	850
CT	650
Nuclear plant	2000
Hydro	1500
Wind	2000

Table 2.6 Technology Characteristics

Technology	Output in MW	Heat rate MMBTU/MWH
Coal	100–1300	10–15
Nuclear	900–1300	10–15
Gas combustion turbine	25–200	9.5–12.5
Gas combined cycle	400–600	6.8–7.5
Traditional gas and oil	25–500	10–15
Wind	0.5–1.5	NA

NA, not applicable.

Table 2.7 Technology Report Card

Technology	Cost of construction	Cost of operations	Reliability	Flexibility	Certainty of fuel supply	Emissions	Other environmental
Coal	C	B	A	B, C	A	D	E
CCGT	B	C	A	B, C	C	B	B
CT	A	D	B	B	C	C	B
Nuclear	E	C	A	B, C, D	B	A	D
Hydro	C	A	C	A	NA	A	B
Wind	E	B	E	E	NA	A	A

Remarks

Below are explanatory remarks for some of the grades.

- Wind, Reliability, Grade E: We rate this very poorly because of additional capacity reserves that must be constructed as a backup for wind generation. Recall the weather impact on load. It is unfortunate that in some areas weather patterns associated with high load also bring no winds (high pressure, high heat systems). Backup reserves for wind systems must be in place to provide reliability.
- Wind, Cost of Construction, Grade E: Aside from installation of windmills there are additional costs. Transmission upgrades to stabilize output are necessary. The need for additional reserve margins also increases system costs. This is a cost that is not reflected in the cost table.
- Cost of Operations: For coal, CCGT, and CT technologies these costs primarily reflect fuel costs. As of the writing of this book coal costs on the order of \$2.50 per MMBTU and gas costs on the order of \$7.50 per MMBTU. Other costs of operations include maintenance and replacement charges. These are the dominant charges of the other systems.
- Hydro, Reliability, Grade C: This grade is certainly arguable. On the hardware side, hydro systems have provided reliable power for over 80 years with few maintenance problems. However, uncertainty in weather creates variable output from year to year.
- Flexibility: This is the ability to follow load. Some systems have multiple entries. This indicates that the technology can be configured in different ways with different outcomes. Nuclear technology in the US is incapable of following load. Output is kept fairly constant. However, the French install load-following capabilities on their nuclear facilities. Note that in France nuclear power produces 77 percent of all electricity production.
- Coal, Other Environmental, Grade E: The primary mover for this grade is the environmental damage of mining. Mining activity induces leaching of acid into groundwater and streams.

- Nuclear, Other Environmental, Grade D: Waste storage of nuclear material is the primary concern.
- Hydro, Other Environmental, Grade B: Primary concerns are flooding of scenic areas, silting of reservoirs, altering ecosystems, and effect on salmon and other freshwater fish.
- Wind, Other Environmental, Grade A: Bird lovers would strongly disagree with this grade. Windmills kill many birds. This is one reason that planners are considering off-shore systems.

Chapter 3

The Grid

Every modern economy develops an electricity infrastructure around large-scale generation plants coupled to load centers by transmission systems. Transmission systems not only couple generation with load centers, they also allow geographically dispersed load centers to economize their use of generation resources. A load center with economical surplus generation can use transmission lines to sell its surplus into load centers with less economical supply. Additionally, transmission systems allow for the sharing of reserve capacity across load centers.

The grid is the network of equipment that supplies, controls, and delivers power. Grid operations are fundamental to balancing load and supply, ensuring reliability and delivery of quality power, and reducing the impact of adverse grid conditions. This chapter describes the grid.

In Section 3.1, we begin with fundamental definitions and principles. We revisit load and generation and explain requirements of alternating current systems. Afterwards, Section 3.2 presents grid equipment and usage. In Section 3.3, we present reserve policy for ensuring system reliability. Section 3.4 provides a description of grid configuration; it describes the location and use of grid equipment with respect to load centers. In Section 3.5, we provide some examples of how grid operators respond to various contingencies. Section 3.6 concludes the chapter with a presentation of the events that led to the Northeastern blackout of 2003.

3.1 FUNDAMENTALS: LOAD, GENERATION, AND ALTERNATING CURRENT

Chapter 2's discussion of load and generation is insufficient to develop an understanding of grid operations. That presentation neglects aspects of demand and generation that are critical for defining and meeting load requirements. This section addresses these issues. In particular, this section provides introductory remarks on alternating current and relates the concept to demand and supply. We start by presenting a more

Electric Power Planning for Regulated and Deregulated Markets. By Arthur Mazer

complete description of generation that includes the generation of alternating current. Afterward, a more complete description of load requirements that takes alternating current into account is presented. A summary of facts and concepts is provided at the end of the section. The remainder of the text is accessible with an understanding of the summary. Although calculus is not required to understand the material, there are occasional references that allow for a better appreciation of the material.

3.1.1 Generators Revisited

The description of electricity generation in Chapter 2 neglects aspects that are critical for understanding grid operations. We begin with a more complete description of electric generation.

A generator creates a voltage potential that results from a changing magnetic field. The rotor's electromagnet rotates in a cyclical fashion. As the rotor rotates, each stator perceives changes in the magnetic orientation of the electromagnet. The orientation cycles as the rotor rotates: A magnetic pole is aligned with a stator, the alignment reverses, then the electromagnet returns to its original position and the original alignment is restored. The cycle repeats itself. The changing orientation causes a changing voltage that also cycles from positive to negative back to positive and so forth. The cycling voltage causes the current to cycle directionally.

The graph in Figure 3.1 depicts the output voltage of a generator as a function of time. The zero point is arbitrary; we assign the point of maximum voltage at *time 0*. A complete cycle occurs when the voltage returns to its maximum value. For a generator with a single set of poles, a complete cycle corresponds to a complete rotation of the rotor. In the US the time between cycles is one sixtieth of a second. The frequency is the number of cycles in a given time frame, normally taken to be a second. This is the reciprocal of the time between cycles; in the US the standard

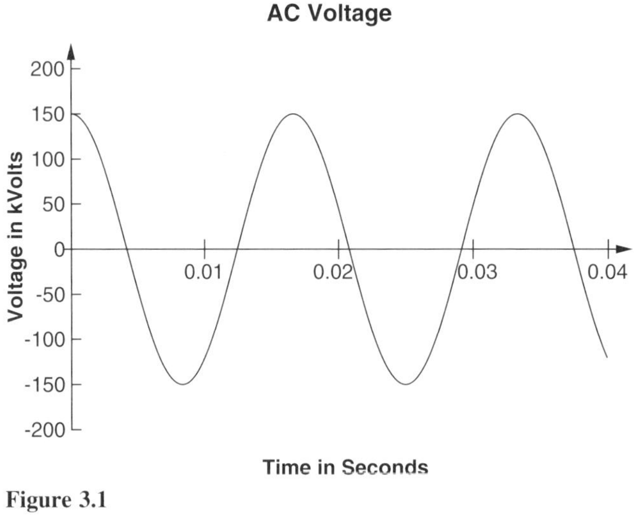

Figure 3.1

frequency is 60 cycles per second. (This is why generators with a single set of poles must rotate at 60 spins per second.) The equation associated with the graph is the following.

$$V(t) - V_{\max} \cos \omega t$$

- $V_{\max}$ is the maximum voltage.
- $\omega = 2\pi f = 120\pi$; f is the frequency of the system, 60 cycles per second.

While not depicted, current cycles along with the voltage, although the peaks and troughs are not generally aligned. Because the current cycles, it is called alternating current (AC).

3.1.2 Load Revisited: Frequency, Voltage, Real and Reactive Demand

The description of load provided in Chapter 2 is incomplete; there is only a brief mention of frequency and voltage requirements, and the characteristics of load in the presence of alternating current are not considered. It turns out that these properties are critical for load requirements and are central to the design and operation of the grid. This section provides a brief overview of these factors.

Frequency and Voltage

The frequency of a power source is critical for any electronic device that has electronic timing elements. Clocks are an example. Other consumer devices include computers, VCRs, DVD players, video games, and microwave ovens. Additionally, there are a host of industrial machines requiring appropriate frequency input, robotics, assembly lines, and many more. Even devices that run on direct current first pass electricity through a transformer that converts alternating current to direct current. In the US, converters are designed to convert current with a specified frequency of 60 cycles per second. Suffice it to say that every residential and industrial customer requires that the power system deliver AC power at the correct frequency of 60 cycles per second.

Voltage requirements are equally critical; appliances and machines do not operate with insufficient voltage, and excessive voltage renders many devices permanently inoperable. The transmission system must be operated so as to ensure that power is delivered to customers with the correct frequency and voltage.

Real and Reactive Power

The power requirements of load are more complicated than indicated in Chapter 2. There are two components to power, real and reactive. Each component is important because each component must be balanced by supply. Indeed, bringing load and supply into balance economically, the central concern of the power delivery chain, requires that power output must match both the real and reactive components of load. In Section 3.4, it is shown that reactive power takes on additional significance in the context of grid operations. A description of real and reactive power follows.

There are two basic types of load elements, resistive elements and reactive elements. Electric machinery and appliances can be described in terms of these elements. Real power is power that resistive load elements require. Reactive power is power that is induced by reactive elements. A qualitative description of these elements follows, with some quantitative relations. We begin with a description of resistive elements.

As their name implies, resistors resist the flow of current. The degree of resistance is measured in a unit called an ohm and indicated by R. In a system with AC, power requirements through a resistor oscillate around a positive fixed value. Below is an equation that describes the power consumption of a resistor within a system powered by an AC generator. (We assume no circuit elements other than the resistor and the generator.)

$$\text{Power}(t) = \frac{V_{\text{max}}^2(1+\cos(2\omega t))}{2R}$$

where

- V_{max} is the maximum voltage drop across the resistor.
- R is the resistance in ohms.
- $\omega = 2\pi f$; f is the frequency (60 cycles/second).
- t is the time variable in seconds.

Figure 3.2 provides a graph of the power consumption in a resistor. The energy consumed by the resistor as a function of time is given by the following equation:

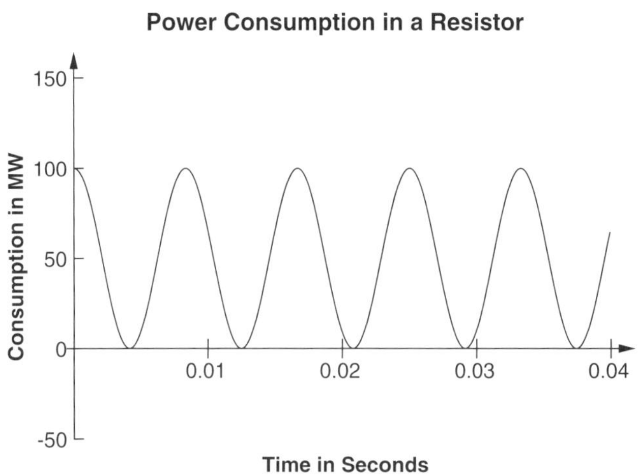

Figure 3.2

$$\text{Energy}(t) = \frac{V_{\text{max}}^2}{2R}\left(t + \frac{\sin(2\omega t)}{2\omega}\right)$$

where all variables have definitions as in the preceding equation.

Figure 3.3 provides a graph of the energy equation. Note that the energy consumed by the resistor is always positive and the trend is to increase as time goes on. Because the frequency, ω, is very large, the term, $V_{\text{max}}^2 \dfrac{\sin(2\omega t)}{2\omega}$, is insignificant in comparison with the term, $\dfrac{V_{\text{max}}^2 t}{2R}$. Accordingly, the consumption is nearly linear in time and becomes substantial with time. Real power is the power from resistive elements that requires energy consumption. Finally, we note that the energy is the area between the time axis and the power axis of the power graph.

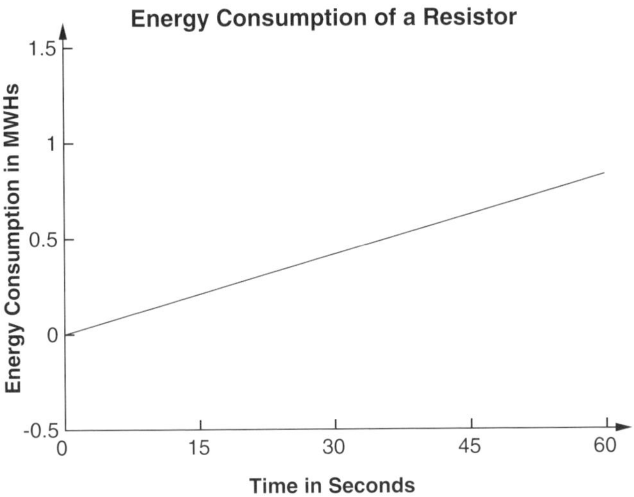

Figure 3.3

Reactive elements are so called because they react to voltage and current fluctuation of AC power. Reactive elements convert and store electrical energy from the system as well as returning the stored energy to the system. As seen below, within AC systems, reactive elements store and return energy in a cyclic manner. There are two types of reactive elements, inductors and capacitors. The power requirement of reactive elements oscillates around zero. The equations of power consumption for inductors and capacitors within a system powered by an AC generator are provided below. (We assume no circuit elements other than the generator and the inductor or capacitor.)

Inductor	Capacitor
$\text{Power}_L(t) = \dfrac{V_{\text{max}}^2 \sin(2\omega t)}{2L\omega}$	$\text{Power}_C(t) = -\dfrac{C\omega V_{\text{max}}^2 \sin(2\omega t)}{2}$

where

- V_{max} is the maximum voltage drop across the resistor.
- R is the resistance in ohms.
- $\omega = 2\pi f$; f is the frequency (60 cycles/second).
- t is the time variable in seconds.
- L is the inductance of the inductor.
- C is the capacitance of the capacitor.

Figure 3.4 provides graphs of the power consumption and corresponds to the equations.

Note the difference between this equation and the equation for the power consumed by the resistor. The power input into the resistor is never negative; the resistor is always consuming power. The power input into reactive elements oscillates from positive to negative, back to positive, and so forth. Whenever the power input is positive, power flows from the generator to the reactive element. Whenever the power input is negative, the reactive element returns power to the system generator. In the graphs the amplitude of the power requirement for the inductor is larger than that of the capacitor. This is solely for the purpose of differentiating the curves.

The energy consumption of inductors and capacitors is the following:

$$\text{Energy}(t) = \frac{-V_{max}\cos(2\omega t)}{4L\omega^2} \qquad \text{Energy}(t) = \frac{CV_{max}\cos(2\omega t)}{4}$$

Figure 3.5 provides graphs of the energy consumption and corresponds to the equations.

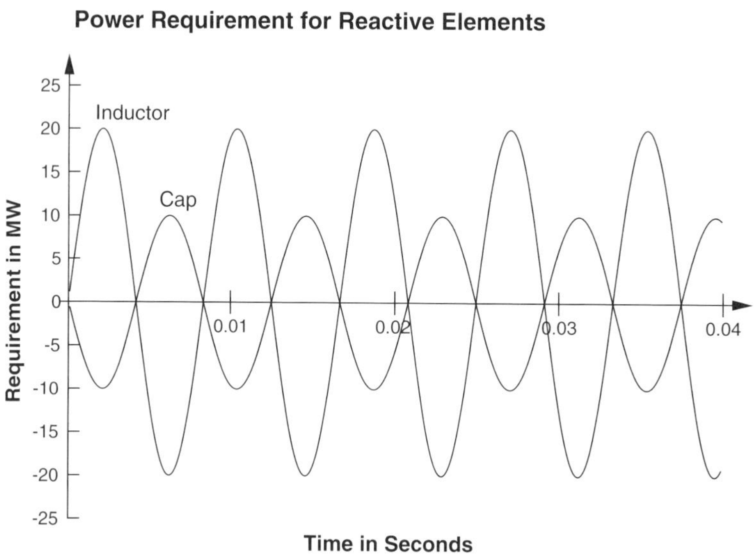

Figure 3.4

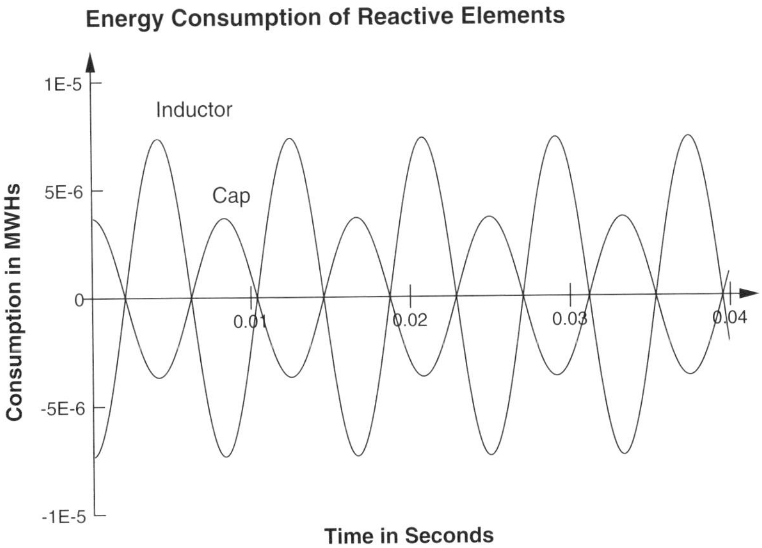

Figure 3.5

Note that the energy associated with the reactive elements oscillates. Unlike the resistor, which continuously consumes energy, a reactive element does not consume energy. The reactive element stores energy and then returns it to the system in a cyclic fashion.

Reactive power is power that remains in the circuit and is not consumed. Even though a net input of energy is not required to supply reactive power, there is an oscillating power requirement with an associated increased current requirement. Power systems must be designed to manage and deliver reactive power. Generators, transmission lines, and other grid devices must be able to manage additional current flows incurred by reactive demand. To distinguish reactive power from real power, the units associated with reactive power are commonly expressed as MVAR as opposed to MW (mega volt-ampere reactance vs. MW; recall that 1 VA = 1 W).

One noteworthy point is the relation between the power and energy requirements of inductors and capacitors. Both the power and energy cycles of a capacitor are complementary to those of an inductor. Whenever the inductor is consuming energy the capacitor is delivering energy, and vice versa. Because of the complementary oscillations, with proper sizing a capacitor can deliver the power requirement of the reactive demand of an inductor.

Another point of note is that the magnitude of the reactive energy oscillations is extremely small. Because of the fast cycling of the alternating current, there is little time for reactive elements to store energy before they release it. Although energy requirements are insignificant, power requirements and current oscillations of reactive elements are not insignificant.

Real Power and Reactive Power Demand

All electrical machinery and appliances have resistive and reactive elements. Take, for example, a vacuum cleaner. The resistance leading to real power is rather intuitive. It is not surprising that energy is lost in rotating the brushes that sweep the floor. After all, there is a lot of friction between the carpet and these brushes and there is a lot of movement of air and dirt particles. The friction and movement of air and dirt consume electrical energy, creating the demand for real power.

There is also inductance in the vacuum cleaner. The vacuum cleaner contains an electric motor with coils. The coils act as an inductor; the AC current creates a fluctuating magnetic field within the coils that alternately stores and releases energy. The simultaneous demand of power consumers leads to an aggregate power demand with both real and reactive components.

Summary

- Load has several requirements: frequency requirement, voltage requirement, real power requirement, and reactive power requirement.
- Supply must meet all of these requirements. Quality supply is supply that remains within permissible voltage and frequency limits while matching real and reactive power requirements.
- Frequency is the number of voltage and current cycles per second of delivered power. In the US the frequency is 60 cycles per second.
- Real power is power that is consumed by resistors.
- Reactive power is power that oscillates between components of an electrical system. Energy is not consumed when providing reactive power. However, current does flow and grid equipment must be able to handle increased current flows due to reactive power requirements.
- The two types of reactive elements are inductors and capacitors. The power and energy oscillations induced by an inductor may be offset with the introduction of a capacitor.
- To distinguish between types of power, different unit conventions are used. The conventions are as follows: Real power is expressed in MW. Reactive power is expressed in MVAR.

3.2 GRID EQUIPMENT

The grid encompasses all equipment that supplies, delivers, and consumes power. Grid operations are the activities undertaken to ensure reliable delivery of quality power or limit the impact of adverse conditions. Before discussing grid operations, it is necessary to understand the equipment that is available to grid operators. This section provides some of the main components along with their functionality and some design characteristics. The list is not exhaustive.

3.2.1 Generators

The generator is a component of the grid that serves various grid requirements. Below is a list.

- Voltage maintenance. The generator provides the voltage requirements of the grid. Recall Faraday's discovery that voltage is induced by a changing magnetic field. In fact, the voltage level is proportional to the rate of change of the magnetic field. The exciter controls voltage output by adjusting the strength of the rotor's electromagnet. Another method of controlling voltage would be to increase or decrease the speed at which the rotor spins, but this would have an adverse impact on the system frequency.
- Frequency regulation. The frequency of supply is determined by the rotational speed of the generator. This is controlled by regulating the mechanical power applied to the turbine.
- Provides real power. The generator is the source of real power for the grid system. Control of real power output is a function of balancing the mechanical power input into the turbine with the generator's voltage output. Exciter levels and the fuel burn are coordinated to perform this function.
- Provides reactive power. A key role of generation is to supply reactive power that is required by load. A generator responds to inductive load with complementary power oscillations. This is accomplished by the exciter's response to voltage conditions. In certain locations there are limited reactive power sources near load centers. Units are designated to provide reactive power and deliver little real power.
- Power limit. The maximum power limit applies to the total power output. Total power output results from both real and reactive power output. Because of the total power limit, there is a diametric relation between a generator's real and reactive power capacity: The more real power a generator delivers, the less capacity the generator has to deliver reactive power. Units that are designated to provide reactive power frequently operate at their minimum power output rating so that sufficient reactive power is available to the grid.

3.2.2 Automatic Generation Control

Automatic generation control (AGC) is a feedback control system that allows generation to respond appropriately to fluctuating grid conditions. Fluctuations come from three sources: load, generation, and transmission systems. Load fluctuates as consumer activity changes and as consumers respond to changing weather conditions. Fluctuations in generation exist as units trip. Similarly, fluctuations in transmission systems occur as equipment trips.

As an example of how AGC works, consider the events on an operating unit that follow the loss of supply (a unit trip or transmission line outage). At the onset of the outage, real power flows across the grid deviate from a preset schedule.

Monitoring systems sense the change in power flows and call upon the AGC of specified units to restore power flows to preoutage conditions. The AGC of an operating unit responds through the exciter; the exciter increases the strength of the electromagnet's magnetic field, which in turn increases the voltage output. The rotor responds by slowing down because mechanical power to the turbine has not been increased. This causes the frequency to fall below its target level of 60 cycles per second. AGC receives inputs of frequency measurements and responds by increasing the power input into the turbine to bring the frequency back to 60 cycles per second. As the example demonstrates, AGC balances the magnetic strength of the electromagnet and mechanical power to maintain appropriate voltage and frequency levels.

Effective AGC requires an effective control mechanism for the mechanical power input to the turbine. The most responsive systems are hydroelectric systems and are used for balancing whenever they are available. The second most preferred generation technology to utilize with AGC is a CCGT. This is because it is much easier to control the burn rate of natural gas than to control coal burn rates and nuclear fission rates.

During the daytime many units are set to run at their optimal output level. However, several units are selected to respond to fluctuating grid conditions with AGC.

3.2.3 AC Transmission Lines

Transmission lines are necessary to carry power between generation sources and load centers. Below are several points that are relevant to grid operations.

- Current limitations and resistance. Transmission lines have operating temperature limitations. At high temperature a line sags and may hit ground, trees, or other objects. This must be avoided; indeed, there is a minimum clearance tolerance between transmission lines and surface objects. Another situation to be avoided is irreversible line stretch, also caused by increased temperature. Increased temperature in a line is due to line resistance, which in turn induces real power loss. The power loss is converted into heat, which raises the temperature of the transmission line. The real power loss is expressed as I^2R, where I is the current and R is the resistance of the line. This demonstrates that thermal limitations on the line impose current carrying limitations.
- Inductance and current limitations. Transmission lines act as inductors. Indeed, under normal operating conditions, power oscillations due to inductance can be on the order of 10 times the power loss from resistance. High inductance creates higher current flows as the inductor contributes oscillatory reactive currents that add to the real current. This significantly impacts on the current limitations of a transmission line and other grid components.
- Inductance, reactive load, and voltage collapse. Because of transmission line inductance, a generator is incapable of providing reactive power to a distant

load. Without reactive power support, voltage requirements at the load center may not be within tolerance bands of grid equipment, and in extreme cases it may be impossible to provide power at any voltage level. This leads to a condition known as a voltage collapse.

- Inductance and power transfer capacity. The inductance of a transmission line limits the real power transfer capability of the line.
- Transmission lines and voltages. Transmission of power over large distances requires high-voltage transmission lines. (Power = VI, where V is voltage and I is current. For a given power requirement, increasing voltage reduces current requirement; power losses and inductance are accordingly lowered.) High-voltage lines are more expensive to construct than low-voltage lines. They require higher transmission towers and wider pathways. Pathway rights are a key cost consideration in locations where land is expensive. For this reason, low-voltage lines requiring narrower pathways are often used over short distances and within population centers. Line voltages vary between 60 kV and 750 kV.
- Transmission outages. Unscheduled transmission outages are usually quickly resolved, often within minutes. Outages that last more than a day are quite rare. Outages commonly occur for the following reasons.
  - Insufficient voltage support (reactive demand not met)
  - System instability; discussed further in Section 3.4
  - Violation of current limitations
  - Insufficient ground clearance
  - Contact with a tree
  - Tower collapse
  - Failure of supporting equipment: transformer, capacitor, meters, communications, and other SCADA equipment
  - Lightning strike creating a voltage disturbance
- Scheduled outages. Scheduled outages occur for planned system upgrades, inspections, and maintenance of support equipment.

3.2.4 DC Transmission Lines

There are situations in which power flows are economical over very long distances. Transmission over great distances can be accomplished efficiently with direct current (DC) lines. DC lines do not create inductive demand as AC lines do. This has two benefits. First, the carrying capacity of real power over DC lines is better than that over AC lines. Second, DC lines do not need system elements to provide the reactive demand of the line. As an example of the use of DC lines, surplus hydro generation from the Pacific Northwest serves the Southern California load. A 500-kV DC transmission line with a power rating of 3100 MW runs 845 miles between Oregon and Los Angeles.

3.2.5 Converters

A converter converts AC to DC and vice versa. Converters are used at both ends of DC transmission lines.

3.2.6 Distribution Lines

Distribution lines distribute power to communities. For example, the lines that are suspended along residential streets are distribution lines. The voltage of residential lines typically ranges between 4 and 12 kV. The voltage of distribution lines supplying an industrial customer may be significantly higher. However, the voltage of all distribution lines is substantially less than that of transmission lines.

3.2.7 Transformers

Power flowing from the generator to a customer undergoes a series of voltage changes. Voltages between the generator, transmission lines, distribution system, and finally the customer, are all different. A transformer is a device that changes voltage between incoming and outgoing lines.

3.2.8 Regulating Transformers

A regulating transformer is used in locations where system voltages change from one transmission line to another and regulation of power flows is desired. The regulation transformer changes the voltage magnitude and phase angle between transmission lines. The phase angle determines the alignment of a voltage wave's peak and trough on the incoming side of the transformer with the peak and trough of the voltage wave on the outgoing side of the transformer. The phase angle is critical for controlling power transfers across transmission lines and is further discussed in Section 3.4.

3.2.9 Circuit Breakers

Circuit breakers decouple equipment or lines from the grid. Circuit breakers operate both automatically and manually. A circuit breaker opens automatically in order to protect equipment from adverse operating conditions. For example, a circuit breaker might open after a lightning strike. Circuit breakers also may be controlled manually to allow system operators to decouple grid components as required. For example, circuit breakers are opened manually when a piece of equipment undergoing maintenance must be decoupled from the grid.

3.2.10 Surge Protection Devices

Surge protection refers to protection against a sudden and dramatic system voltage change. Lightning strikes are a frequent cause of voltage surges.

3.2.11 Additional Lightning Protection

There is a wire between transmission towers that acts as a lightning rod. The wire is grounded through the transmission towers.

3.2.12 Reactive Power Providers

Below is a list of devices that provide reactive power. Because of the inability of transmission lines to deliver reactive power over long distances, reactive power sources must be available locally, close to load centers.

- Generator. Generators respond to voltage changes to provide reactive power as demanded by load.
- Capacitor. Capacitors provide reactive power that is complementary to inductive loads. They are often used to provide voltage support over transmission lines.
- Inductors. Inductors provide reactive power that is complementary to capacitive load.
- Static VAR compensators. These devices act as inductors or capacitors with automatic fast switching capability. Switches for inductance or capacitance respond to load conditions to provide quick reactive support.
- Static synchronous compensator (statcoms). Statcoms are solid-state electronic devices that can behave as either inductors or capacitors. Statcoms respond automatically to system needs.
- Servomechanical compensator. These are mechanical devices that are similar to a generator. They have an electromagnet on a rotor as well as stators. The rotor is not coupled to a turbine and does not spin because of power input from off-grid machines. Instead, the stators create an oscillating magnetic field in response to reactive power flows on the grid. In turn, the electromagnet on the rotor responds, causing the rotor to spin. As the rotor spins the compensator generates the reactive power requirements of the grid.

3.2.13 Station Bus

Station buses are points of interconnection between transmission lines, transmission lines and distribution lines, or interconnection points from a generator to a transmission line. Transformers, circuit breakers, meters, and voltage support devices are often collocated at station buses. Station buses are also frequently referred to as substations.

3.2.14 Distribution Switches

Recall that a generator has three stators. The phase angles of the voltages from each stator all differ, meaning that the peaks and troughs of the voltage waves occur at different times. The power from each stator is carried through the grid and delivered to the customer while preserving the differences in the phase angles. The configuration is economical because it is unnecessary to construct return lines to the generator. However, the configuration requires that system loads are evenly distributed among the three phases. At larger distribution stations, there are switching centers that balance load among the three phases. There are also switching devices within the distribution system that balance loads among the three phases.

3.2.15 SCADA and Monitoring Systems

Automatic control systems as well as system operators require information concerning system status. There are voltage and current meters at station buses. There is a communications network to relay information between all of a utility's operating facilities and its control room; AGC is coupled to this network as well as automatic controls for other grid equipment. The technology for monitoring and coupling data with control systems has the acronym SCADA: supervisory control and data acquisition.

3.2.16 Black Start Generator

Recall that starting a unit requires electricity to power the electromagnet on the rotor. This is the function of the exciter. In case of a blackout, electricity from the grid is unavailable and it is not possible to restart a unit without an alternative power supply. The alternative electric supply is known as a black start generator. Typically, a black start generator is a small diesel generator along with a battery pack. The battery pack provides the electricity required to start the diesel generator. The diesel generator in turn provides the electricity necessary to start the larger unit. Not every unit requires black power. In case of a blackout, a limited set of units can restart using black start equipment. These units provide enough power to restart other units on the grid.

3.2.17 Plant to Customer

Figure 3.6 illustrates the delivery of electricity from the power plant to the customer. In the figure, A represents a power station, B a transformer, C a regulation

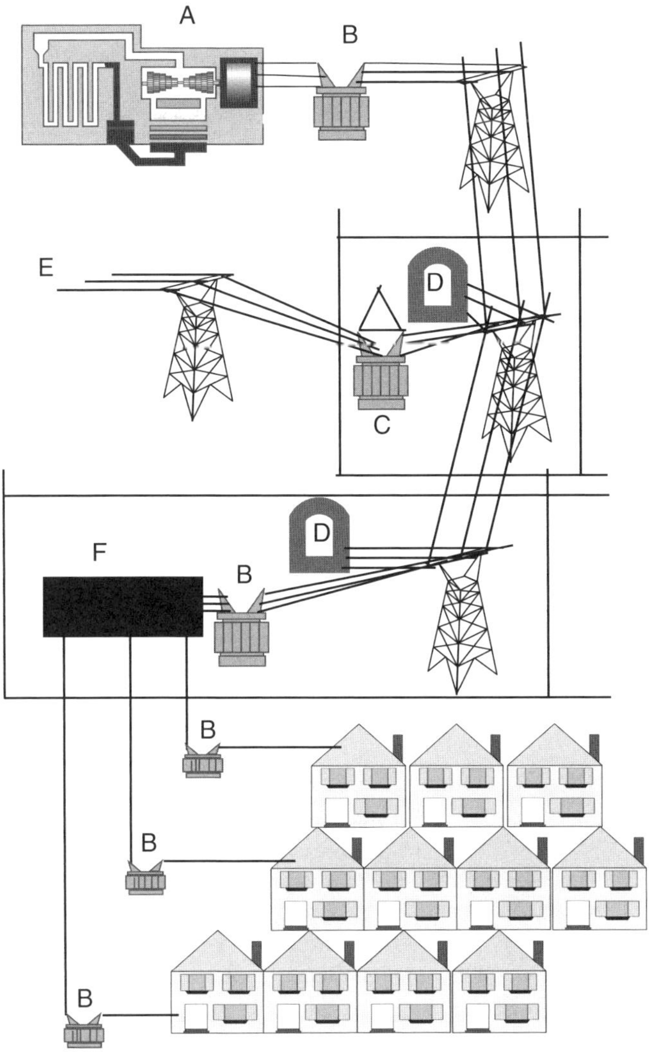

Figure 3.6

transformer, D a reactive power element, E incoming power, and F a distribution switch.

1. Power is generated at the power station. (A)
2. A transformer at the plant alters the plant output voltage to match line voltage. (B)
3. Power flows across a transmission line to a station bus.
   3.1 Voltages across power lines are matched and phase angles controlled with regulation transformers. (C)
   3.2 Devices provide reactive power requirements when necessary. (D)

4. Power flows across transmission lines to a delivery station bus.
   4.1 Devices provide reactive power requirements when necessary. (D)
   4.2 A step-down transformer reduces voltage to match distribution line requirements. (B)
   4.3 Power passes through a distribution center, where the load is balanced among the three phases. Only large distribution stations have associated distribution centers. Most distribution substations do not have distribution centers, and balancing of load among the phases occurs through automated processes after the final step-down transformer. (F)
5. Distribution lines deliver power to customer lines.
6. Another step-down transformer between the distribution wire and the customer line reduces voltage to match customer requirements. For households in the US this is 110 volts. (B)
7. Distribution lines deliver power to end customers.

Remarks

- The three lines from the power plant at A to its associated transformer B each connect to one stator and have their own separate phase. Transmission towers carry lines in sets of three; each line corresponds to a phase.
- Transmission towers carry more than one line to enhance reliability. This provides system redundancy in case one or more lines fail.

3.3 GRID RELIABILITY AND CONTINGENCY REQUIREMENTS

Electricity supply must be generated and delivered to match demand in real time. As noted in Chapter 2, there are temporal operating restrictions that limit the rate at which units can be brought on line and alter their output. These include ramp-up rates and minimum downtimes, among others. Because of temporal restrictions, balancing supply to meet demand in real time requires an advance plan that includes contingency planning. Operational support must be available for responding to changing system conditions in real time. The advance plan is set each day for the following day and includes a schedule for unit operations. Grid operators then execute the plan and adapt it to meet the requirements of changing conditions. This section examines contingency planning and real-time operational requirements that are necessary for balancing supply with load in real time within the operational requirements of the grid. The real-time services necessary to ensure uninterrupted power in the case of changing conditions are known as ancillary services.

In the US, all grid operators are required to execute plans that allow for uninterrupted power delivery in the event of equipment failure. The most common failures involve transmission lines and support equipment. This is because transmission lines are exposed to weather and weather causes outages, for example, lightning strikes. Nevertheless, generation units also fail and contingency plans are developed to address unit outages.

The day-ahead plan is subject to contingency analysis. Simulations of transmission line and unit outages are run through computer models, and the plan must demonstrate its ability to continue service through the outages. Scenarios include outages of equipment within the operational control area as well as neighboring control areas.

Contingency planning requires reserve capacity for both transmission systems and generation resources. Reserve capacity in transmission systems is attained by scheduling power transfers across transmission lines at lower levels than the lines' ratings. In this manner, if a surge in power occurs in a given line because of a failure elsewhere in the grid, the line has the capacity to handle the additional power flow.

The operating rule is that reserve capacity must be restored if an event occurs that eliminates the reserve capacity. For example, suppose that a transmission line fails and circuit breakers open. Suppose that this failure causes power to flow along an alternative transmission line. As noted above, contingency planning should predict the flow patterns after the failure of the original line, and the alternative pathways should have enough spare capacity to handle additional power flows. Suppose further that the postfailure flows are not able to manage another line outage. This situation violates the requirement that the grid must always be able to operate without interruption in the event of equipment failure. In this case, grid operators must take corrective action so that the grid could once again withstand an equipment failure. Corrective actions are discussed in Section 3.5.

Generation reserves are required to handle contingency outages. Policy for reserves is set at a regional level by regional reliability councils. There are eight reliability councils that encompass the US, Canada, and Mexico. Coordination of policy across reliability councils occurs at the federal level in the Federal Energy Regulatory Commission (FERC) and the corresponding agencies of Canada and Mexico. The policies vary between and within reliability councils; however, the policies have common features. There are several levels of reserve requirements: primary reserve, secondary reserve, and additional reserves.

The primary reserve has two components, spinning and quick start. Spinning reserves are resources that are synchronized to the grid and can attain their spinning capacity within 10 minutes of being called upon. Spinning reserves are available for meeting demand in the event of a unit outage, a transmission outage that restricts imports, or load swings that are not manageable by regulating capacity (discussed below).

Quick-start reserves are those that are available to come on line and attain their spinning capacity within 10 minutes. These reserves include diesel-fired power generation and combustion turbines. In contrast to spinning reserves, quick-start reserves do not have to be synchronized to the grid. Note that in some regions, contractually interruptible service qualifies as meeting the quick-start reserve; that is, demand reduction through contractually agreed-upon arrangements with customers provides the quick-start reserve.

The primary reserve requirement is set according to the principle that systems must operate in a manner that ensures uninterrupted service even in the event of equipment failure. According to this principle, primary reserves must be sufficient to replace power for the largest single point of failure. Typically, this is the largest unit within the control area. For example, if a 1150-MW nuclear power plant is the largest unit within a given control area, 1150 MW of primary reserves are required to maintain uninterrupted service in the event of a unit outage. The percentage of spinning reserves that accounts for primary reserves varies in accordance with import capabilities and grid configuration. However, it is no less than 50 percent of all primary reserves and in some cases 100 percent of the primary reserves.

The operating rule that requires restoration of reserves applies to primary reserves. Continuing the above example, suppose the nuclear unit trips and 1150 MW of primary reserves are called upon to deliver power. Suppose that the next largest unit is a 900-MW coal plant. Then another set of 900 MW must be made available as primary reserves, and the spinning component of the new primary reserve set must be greater than the minimum allowable percentage.

Secondary reserves are idle reserves that are available to replace the primary reserves. Secondary reserves must be able to come on line within a prescribed time frame. The time frame differs among control areas but is normally between 30 and 60 minutes. The capacity requirement for secondary reserves varies among control areas. Additional reserves are those that take longer than the prescribed time to come on line. Control areas maintain a capacity base that is based on their forecasted peak load requirement. A common standard is to target around 115 percent of the forecasted peak load. After accounting for dispatched units and primary and secondary reserves, the additional reserves are those that meet the installed capacity target.

Black start capability that allows a unit to restart in the event of a blackout is also a component of contingency planning. Plans are drawn up to allow a coordinated response to a blackout. Units with black start capability must first restart to bring power back to the grid. Once black start units are brought on line and grid power is available for the exciters of units without black start capability, those units can also be brought back on line. Engineers must determine the number of black start units that are necessary to perform this task. Contingency planners have a specified sequence for restarting units in the event of a blackout.

Load-following capabilities are necessary to ensure service reliability. As noted in Section 3.2, most units are set to run at their optimal efficiency once they have come on line and ramped up. However, a specified set of units are identified as system regulation (load following) units; these units have AGC capability. They must have satisfactory ramp-up and ramp-down rates to match load fluctuations. Hydro units are the preferred units for providing load-following capability because of their quick ramp-up and ramp-down capability. The number of MW required for load-following services is a percentage of the forecasted load that varies among control areas and falls within the range of one to two percent of the forecasted load. Load-following services are frequently called regulation services.

3.4 GRID CONFIGURATION

This section provides an overview of the configuration of a network. There are five functions that the grid serves.

1. Provide real power requirements.
2. Provide reactive power requirements.
3. Provide system control.
4. Ensure system reliability.
5. Ensure system security.

The overview refers to the schematic of a network for a single control area as provided in Figure 3.7. The schematic provides the location and size of grid equipment. We identify the equipment that supports grid services and requirements.

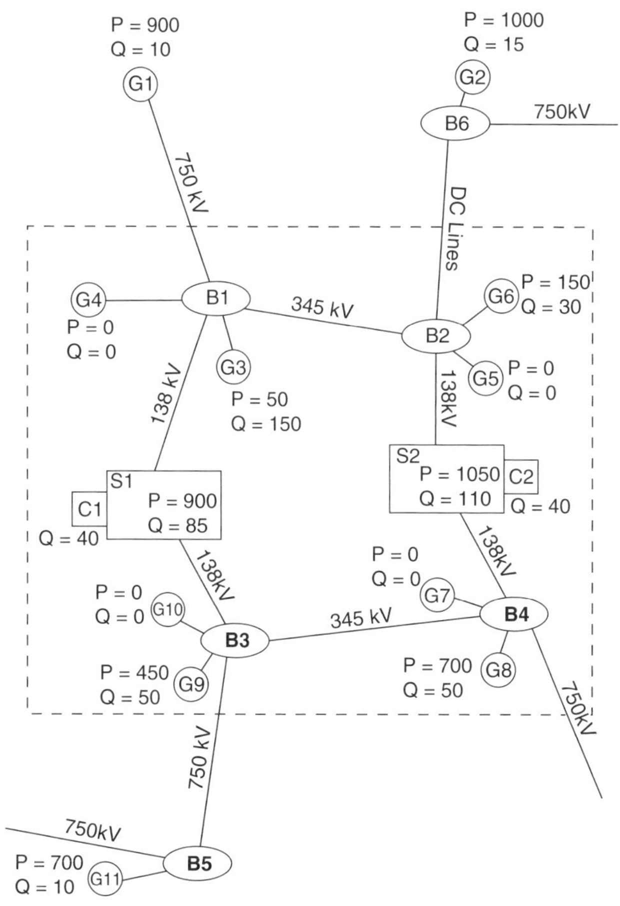

Figure 3.7

3.4.1 Provide Real Power Requirements

A central function of the grid is to provide real power requirements of the load. The real power loads at the substations are indicated by the letter P. For example, at substation S1 the real power requirement is 900 MW.

In the schematic in Figure 3.7 there are eight generators within the control area and three generators outside the control area. The control area is the region within the dotted lines. The real power generated by each generator is indicated by the accompanying letter P. In addition to the generators and distribution buses within the schematic, there are generators and load centers in neighboring control areas that influence the delivery system but are not a part of the schematic.

The transmission system is necessary for providing real power to load. Transmission systems must be designed to perform this efficiently. There are DC transmission lines from generator G2's interconnection bus, a substantial distance from the control area, to bus B4 within the control area. As previously noted, DC lines are superior for transmitting power over long distances. DC lines do not require reactive power support and DC lines have higher power transfer capacity than AC lines.

There are converters (not shown) on the delivery and receiving sides of the DC lines. On the delivery side, a converter converts power from AC to DC, while on the receiving side at bus B2 a converter converts DC to AC. In the schematic, not all the power from generator G2 flows across the DC lines. A portion of the power flows to another control area across the 750-kV lines that interconnect with bus B2.

Power from the distant generators, G1 and G11, is supplied via 750-kV lines. Note that these are the highest-rated AC lines in the system. High-voltage lines are more efficient than low-voltage lines because for a given real power transfer, the reactive inductance of a high-voltage transmission line is lower than the reactive inductance of a low-voltage power line. As noted above, reactive inductance in the power line limits the efficiency of power transfer and requires reactive power providers.

Aside from the distant generators there are several generators within the control area. The ratings of the transmission lines that couple these generators with the substations are lower than those of the 750-kV lines that connect generators from outside the control area. While higher-voltage lines are always superior in transmission efficiency, it is not always feasible to construct high-voltage lines. Pathways for high-voltage lines must be much wider than those for low-voltage lines. Property prices for pathway rights are much more expensive in populated areas, where the load centers are located. As such, it may not be economically feasible to purchase the broad pathways necessary for high-voltage lines, and lower-voltage lines are constructed instead.

Distribution lines (not shown) from the substations to the end customer have the lowest kV rating of all lines, 4 kV to 12 kV for residential lines. Distribution lines leading to industrial customers may have higher voltage.

Transformers (not shown) are required to transform the voltage from one standard to another as power passes through transmission lines with different ratings.

Indeed, there is a transformer at each customer location (every household) reducing the voltage of power from the distribution line to customer requirements, 110 V for households.

Note that the sum of the real power output from all generators is greater than the sum of the real power requirement at the substations. This is because of power losses through the transmission system. There are further losses through the distribution system. Real power losses from generator to residential customers may be on the order of magnitude of 10 percent of the total generation. The balance law for real power is that the real power generated is equal to the sum of the real power of the load requirement and the real power losses through the transmission and distribution system.

3.4.2 Provide Reactive Power Requirements

There are two components of the reactive power requirement. The first component is the reactive demand of the customer. The second component is the reactive demand of the transmission network. In the schematic in Figure 3.7, the reactive supply and demand is represented by the symbol Q. The reactive demand of load at the substation reflects reactive demand from customer equipment as well as the distribution system, but not the transmission lines. Note that the reactive production from generation and other reactive power providers exceeds the reactive demand of the load. This is because much of the reactive generation supplies the reactive requirements of transmission lines.

There are several challenging aspects to providing reactive power. First, because of high reactive losses through the transmission system, it is not economical to provide reactive power from distant generation sources; as such power systems are not designed to transmit reactive power over great distances. In the schematic, the generators G1, G2, and G11 provide negligible reactive power for meeting reactive demand at the substations. Conversely, generators close to the load centers, G3, G8, and G9, are the largest source of reactive power. Another challenge of delivering reactive power is that the reactive demand of the transmission system must be met or the system will not function; a transmission line will not transmit either real or reactive power if the reactive demand of the line itself is not satisfied. This is particularly critical for lengthy transmission lines, as reactive losses through a transmission line increase with length. In the schematic, the generator G3 interconnects into the same bus as the distant generator G1. The generator G3 runs at minimum real power levels so that it is able to deliver considerable reactive power. There are also series of capacitors (not shown) along transmission lines known as shunt capacitors. These provide additional reactive power to meet a line's inductive demand.

Yet another challenging aspect of reactive power is that reactive power requirements are less steady than real power demand. To meet uncertain surges in reactive demand, a system of reactive demand providers as identified in the previous section are deployed throughout the network. In the schematic, capacitor banks C1 and C2 are located at the distribution stations S1 and S2 and are available to meet demand fluctuations. The generator running at minimum power is also capable of supplying large swings in reactive demand.

3.4.3 Provide System Control

System control is accomplished both automatically and through human intervention. In either case system control has two components, a monitoring component and a response component. Monitoring of current through transmission wires and voltages at the buses provides a set of information to response systems. In return, response systems activate controls in accordance with operating protocols.

There are several types of control devices. These include controls on plant output and frequency, controls that direct the power transfer across transmission lines, and controls that can segregate equipment from the rest of the grid to maintain system security.

Controls on power plant output and frequency may be coupled to the network through SCADA and AGC to respond to network fluctuations. Alternatively, these controls may be set locally to maintain steady output. In the schematic in Figure 3.7, generators G1 and G8 are set to maintain a constant output. These are baseload units that provide low-cost power. Generators G6 and G9 are coupled with AGC, and their outputs respond to load and supply fluctuations.

It is not possible to direct power flows across the network from point to point. For example, it is not possible to direct all the power from G9 toward distribution station S1. Power will flow along the path of least resistance in accordance with physical laws. In general, the power from generator G9 will flow along all transmission lines emerging from bus B3 where it is interconnected. The allocation of the power transfer along the pathways depends on the state of the entire network, including that of neighboring control areas not visible in the schematic. The state of the network is not constant; accordingly, the allocation of power flows from generator G3 changes with time even if the output from the generator is constant.

Although it is not possible to direct flows from a generator to a specified destination, there is limited capability to control power transfers across transmission lines. Power transfers depend on voltage differences between buses: both the difference in magnitude of the voltage and the difference in the timing of the voltage cycles. The power flow across the transmission lines between buses B3 and B5 depends on voltage differences at those buses. The voltage at bus B3 is determined predominantly by generator G9, while the voltage at bus B5 is determined predominantly by generator G11. Differences in the timing of the voltage cycle can be controlled by the adjusting the difference in the rotor positions of the units. One can increase power flows across the transmission line by increasing the difference in the rotor positions. This provides limited control because one cannot increase the differences in rotor positions too much without causing system instability; system stability is addressed below.

Another way to control the timing of the voltage cycles is by a phase shifter. The regulating transformer at bus B3 can alter the voltage cycle of power flowing across the transmission lines between buses B3 and B5. This allows for control of power transfers across the transmission lines without adjustments to rotor positions that may cause instabilities. In the schematic, the system is set to transfer a specified power level from bus B5 to bus B3. Other transfers may include a flow of power out of the control zone along the 450-kV pathway emanating from bus B4.

Another control feature of the grid is the ability to disconnect equipment from the grid. This feature is important for grid security as discussed below. Grid equipment is isolated by a series of circuit breakers. Circuit breakers (not shown) are activated whenever monitoring systems detect potentially debilitating circumstances. For example, if the current on the transmission line running between buses B3 and B4 exceeds the current tolerance of the transmission line, system monitors detect the current overflow and activate circuit breakers at either end of the transmission line.

3.4.4 Ensure System Adequacy

System adequacy is the capability of maintaining delivery of power for aggregate load throughout established outage scenarios. Adequacy requires redundancy of both generation and transmission equipment as described in Section 3.3. Adequacy refers to the availability of real and reactive power requirements as well as sufficient transmission capacity to withstand contingencies.

In the schematic in Figure 3.7, generators G3, G6, G8, and G9 are on line, but do not operate at their maximum output. They are capable of providing spinning reserves. Generators G10 and G4 are off line, but capable of coming on line within 10 minutes. These generators provide quick-start reserves. Finally, generators G5 and G7 provide secondary reserves.

Power flows over the transmission lines are simulated and analyzed before a final dispatch schedule is set. This activity is further described in Chapter 4. The analysis must show that the current through each transmission line is below the transmission line's current carrying limit. Furthermore, as indicated in Section 3.3, the transmission system must be able to withstand a set of contingencies. Should there be a disturbance to the system causing an alteration of the power flows, the current flow through each transmission line would be below its current carrying limit after the disturbance.

In the schematic, there are multiple lines associated with each pathway. This provides redundancy that allows for line outage contingencies. Additional redundancy occurs as each distribution station is connected to generation through multiple pathways.

Aside from equipment redundancy, correct operations are required to ensure system adequacy. For example, it is noted above that one can control the transfer of power between two buses by adjusting the phases of the voltage cycles between the buses. Larger differences in the phase allow for greater power flows. However, if the difference becomes too large, the system becomes unstable. A small perturbation to the system causes a failure across the transmission path, and power cannot flow.

Frequency regulation is compromised and units trip off line. Outages across one transmission path may have wider system implications.

3.4.5 Ensure System Security

System security refers to the grid's ability to withstand sudden and potentially extreme disturbances such as short circuits or the loss of a major system component. Withstanding a disturbance does not necessarily imply ensuring uninterrupted delivery. It does include precautionary measures that protect against injuries and damage to equipment. The area is very broad, encompassing government regulation that applies to specific types of equipment. As an example of equipment that provides system security, there is a network of circuit breakers (not shown) that allows for the isolation of equipment from the grid; isolation occurs when circuit breakers open. The circuit breakers open in the event of an equipment failure or a disturbance to the grid such as a weather-related transmission tower collapse.

3.5 GRID OPERATIONS

Grid operations are the coordinated activities that result in the balancing of supply and demand. The activities include contingency planning, monitoring, and operating. This section gives a brief description of grid operations. It includes a discussion of monitoring grid operations and coordinating responses across control areas. We then give some examples of operational responses to system events.

3.5.1 Contingency Planning

In Chapter 4, the setting of the day-ahead dispatch schedule is presented. Contingency planning ensures that unit schedules and the corresponding power flows through the transmission system are able to deliver required power to meet load demand under various contingency scenarios. Contingency plans address sudden events that impact the grid.

Events with immediate impact on grid operations include large unit trips and transmission system failures. Within the first 30 seconds of such an event the concern is system instability. The grid becomes unstable when the phase angle between units begins to fluctuate in a random way. This may occur if the exciter of a unit or several units responds to the outage by increasing the electromagnet's magnetic field but there is a slow response in the fuel burn rate. The demand for increased voltage output without a corresponding increase in the fuel burn rate slows down the rotor, and the phase angles of the units fluctuate randomly. Power transfer is not feasible under such circumstances, and there is a voltage collapse at the buses where units interconnect. If the system does not stabilize, a blackout may result. Systems are designed to withstand an outage of significant proportions, such as a nuclear outage, while maintaining system stability. Planning for such events occurs in the design phase, where studies that simulate system outages lead to design requirements.

Once the grid withstands the first 30 seconds of an event and maintains stability, the next concern is the sufficiency of power to maintain operating voltages and meet demand. Providing reactive power is of great concern when a local unit trips off line. Both spinning reserves and reactive power providers are necessary to maintain voltage over a 10-minute time frame, the time it takes to bring remaining primary reserves on line. Spinning reserve requirements and grid facilities for maintaining power during an event are determined well in advance of the event by simulation studies.

Once sufficient power is restored to the grid, the next concern is that the postevent operating conditions are within operational tolerances of all grid equipment. If the operating conditions are within operational tolerances the conditions are said to be feasible. Otherwise, the conditions are infeasible. Simulations are performed to determine the feasibility of various operating conditions. First load, transmission, and dispatch scenarios are developed. Then the power flow through the grid is simulated for the various scenarios. The problem of simulating and solving for the power flow is known as the power flow problem. The solution to the power flow problem is feasible or infeasible depending on whether or not the outcome satisfies system operating constraints, that is, current limitations on transmission lines. As with the stability analysis, simulation studies are performed well in advance of an event to ensure that postevent operating conditions are feasible under various operational scenarios.

The problem of ensuring feasible operating conditions is one that is addressed on a daily basis. As mentioned above, planners establish a dispatch schedule a day ahead of delivery time. The schedule is subject to a power flow analysis to ensure feasibility of the schedule under various operating assumptions. The day-ahead plan is updated right through to delivery time as grid conditions change. Modern computational capabilities allow for the solution to the power flow problem in reasonable time frames. It is possible to solve the power flow problem several times between the setting of the day-ahead schedule and delivery time to test the feasibility of updated plans. The power flow problem is intimately connected to dispatch planning. Accordingly, unit dispatch and transmission planning is an integrated process that cannot be separated.

Contingency planning includes an escalating series of interventions to maintain service, minimize out of service times, or reduce the consequences of adverse conditions. Planners must identify conditions in which to exercise the interventions. Adverse conditions are caused by excessive load, unit outages, or transmission system outages. Potential interventions include the following.

- Bringing more units on line to provide additional real power
- Increasing imports to provide additional real power
- Decreasing exports to provide additional real and reactive power
- Bringing more local units on line to provide reactive power
- Interrupting the service of customers that have negotiated cheaper rates in exchange for the risk of having service interruptions (interruptible load)

- Executing rotating blackouts. Rotating blackouts are blackouts that limit the time duration for the affected customers by rotating power blackouts among different communities.
- Executing geographically limited blackouts. These are blackouts that sacrifice service to a specified locality in order to preserve service elsewhere. The large-scale blackout of 2003 could have been confined to northeastern Ohio had appropriate actions been taken.

3.5.2 Monitoring

In this subsection we give a bottom-up view of monitoring the grid and coordinating grid operations. Monitoring of the grid occurs at different levels. The lowest level of monitoring is the control area. In a traditional utility environment, each utility is responsible for monitoring the grid within its control area and following regulations that are set by regional reliability councils. There are regional reliability coordinators that often encompass the control areas of multiple utilities and monitor utility operations within their territory. Reliability coordinators do not own independent monitoring systems. They receive system information from utilities within their territory. While utilities monitor and operate their own systems, they are obliged to respond to orders from their reliability coordinator. Figure 3.8 provides a map that indicates the different reliability coordinators within the US.

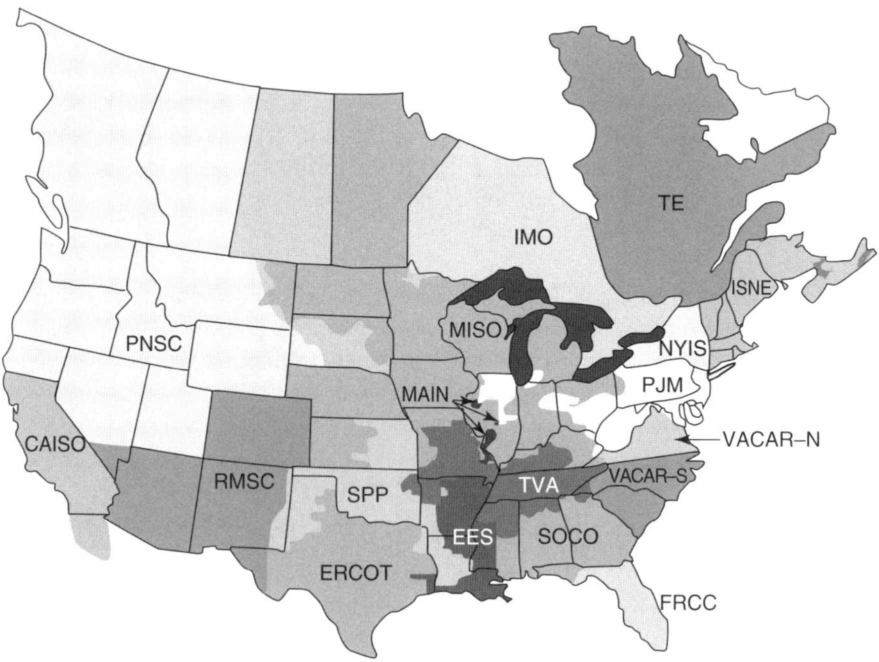

Figure 3.8

Within each utility is a control group responsible for monitoring the operations of the utility's control area. The control group monitors current, frequency, and voltages across transmission lines, bus voltages, and the status of generation units and power transfers within its territory. Anomalies are observed and corrected through automated systems or reported to a decision making authority within the control group. This authority may decide on an appropriate corrective action.

The control group passes system information on to the reliability coordinator. Whenever an event impacting more than one control area occurs, control groups must coordinate with one another, often through the reliability coordinator.

3.5.3 Operations

This subsection provides examples of common control group responsibilities as well as responses to frequently encountered events. Each facility on the grid has either personnel on site to operate equipment or remote operational capabilities. Personnel receive operating instructions from their corresponding control group and execute the instructions accordingly. This list is an addition to the list of potential contingency planning actions presented above.

- Regulation. Control groups are responsible for ensuring that regulation capabilities are available at delivery time. Reserves set aside for regulation are able to accomplish this with AGC. If variations from forecasted load become too high, spinning reserve capacity may be used to follow load. Under this condition additional units must be brought on line to maintain the spinning reserve requirement.
- Unit outage with moderate impact. Control groups must respond appropriately to unit outages. The impact and response to unit outages vary greatly depending on the status of the system as well as the location and size of the unit. In the best case, spinning reserves are able to take up the load and new spinning reserves are brought on line. The events occur as follows.
  1. Units within the control area respond by increasing their output with AGC. This includes spinning reserves.
  2. Imports from outside of the control area automatically increase as power flows respond to the additional power requirement. (Units outside the control area also increase their outputs automatically.)
  3. Capacitors and other reactive power-providing devices respond to an increased need for reactive power. There are automatic systems as well as systems that require operator intervention under the direction of the control group.
  4. The control group informs neighboring control groups and the reliability coordinator to coordinate responses.
  5. In severe cases, the control group may shed interruptible load and implement rotating blackouts.
  6. The control group establishes a new set of primary and secondary reserves.

7. A new dispatch and export/import schedule is set to economically meet load in the absence of the tripped unit. Export/import schedule adjustments are coordinated with neighboring control groups.

- ACE is moderately outside its allowable levels. ACE stands for area control error. This is the difference between scheduled imports/exports and actual imports/exports with neighboring control areas. It is the responsibility of the control group to monitor ACE and maintain the error within allowable limits. When the error is moderate, the error may be corrected by adjusting the phase angle of the voltage on the importing or exporting bus to allow for an appropriate adjustment of power flows.

During an event that severely impacts reliability such as a unit trip or transmission outage, ACE may be ignored as schedules are readjusted to accommodate reliability requirements.

- Import disruption. An import disruption may occur because of a failure of a transmission line or a unit trip within a neighboring control area. In such cases the importing control area responds in a manner similar to that for a unit outage.
- Transmission line outage. The impact of a transmission line outage may be severe or limited. Under normal operating conditions, the grid can maintain service reliability because of the maintenance of transmission reserve margins. Alternate pathways are available to maintain power flow schedules between control areas. More severe outages disrupt the flow of power from an exporting control area to an importing control area. The importer responds to the event as if it were a unit outage. The exporter responds by reducing the output of generation within its control area.

Remark

- Regional coordinators do not set policy; the regional reliability council assumes this role.

3.6 BLACKOUT AUGUST 14, 2003

This section provides an overview of the events that led to the 2003 blackout that affected areas of Ohio, Michigan, Pennsylvania, New York, and Ontario. The section illustrates several points of interest:

- This case illustrates the coordination of various agencies in order to maintain grid reliability (although such coordination was ineffective in this case).
- This case shows the extent to which the transmission network is utilized.
- This case demonstrates how important it is to execute mundane operations properly.

The events that led to the blackout occurred in the control area of First Energy. First Energy is a regulated utility primarily servicing northeast Ohio. The major metropolitan area in First Energy's territory is Cleveland. First Energy is responsible for grid operations in its territory and is under the jurisdiction of the Midwest Independent System Operator (MISO).

MISO is the reliability coordinator for 35 control areas over a region that spans several states including all or parts of Ohio, Indiana, Illinois, Missouri, Michigan, Wisconsin, Iowa, South Dakota, and North Dakota. All of these states are part of the Eastern Interconnect, a series of interconnected control areas that runs from the East Coast to the midwestern states. As coordinating agent, MISO monitors the grid across its territory, including flows into and out of MISO. MISO also coordinates with other reliability coordinators in the Eastern Interconnect. While each utility within MISO's territory is responsible for operating the grid within its control area, MISO has the authority to order appropriate actions for the purpose of ensuring grid reliability.

First Energy monitors, plans, and controls its system through its control center. On the day of the blackout the control area was experiencing difficulties that are typically encountered on high-load days. High temperatures caused higher than normal load demand. Higher than normal system imports were scheduled to meet this demand. The imports caused voltage support problems as local generation was inadequate to meet reactive power demands. The voltage support problems were exacerbated by the forced outage of First Energy's East Lake Unit number 5 in the Cleveland area. The outage removed 600 MW of capacity from the system.

These problems by themselves were manageable. The system had shown itself capable of supplying higher load levels during the previous year. However, additional problems with the transmission system were not appropriately addressed. The overriding reason for not addressing transmission problems was that First Energy's monitoring system had failed. Indeed, at 2:32 P.M., a system operator from a neighboring utility, AEP, had telephoned First Energy's control center to confirm a temporary trip of a 345-kV transmission line that runs between the two utilities' control areas, the Star-South Canton line. First Energy was unaware of the trip, indicating a system monitoring failure. Without correct information, planners at the control center could not respond appropriately to a series of transmission line failures that eventually cascaded into the blackout.

MISO was also unaware of First Energy's system status. MISO received its information directly from control areas within its territories. The failure of First Energy's monitoring system meant that First Energy passed incorrect system status information on to MISO. Without correct system status information, MISO was unaware of the severity of problems in Ohio and could not execute its role as reliability coordinator.

Three 345-kV transmission lines within First Energy's control area tripped without returning to service between 3:06 P.M. and 3:41 P.M. because of contact with ground vegetation. These were the Chamberlain-Harding line, the Hanna-Juniper line, and the Star-South Canton line. Additionally, the Stuart-Atlanta 345-kV line tripped at 3:32 P.M. Power flows responded to transmission outages by finding alternative paths. The outages placed the network in a precarious state. First Energy, MISO, and the operators in First Energy's neighboring control areas were unaware of the severity of the conditions.

The loss of the 345-kV power lines caused the collapse of First Energy's network of 138-kV transmission lines in a cascading wave of line outages. A cascading wave comes about as a result of lines overloading because of a compromised system. Power flows responded to outages by finding alternative pathways. Whenever the flows alter, some lines become overloaded beyond their current limits, causing them to trip. These trips cause subsequent power flow shifts, which once again overloads some lines and once again causes another set of line trips. The trips expand throughout the network as a wave. The cascading wave of 138-kV lines began at 3:41 P.M. and ended with much of Ohio in blackout conditions at 4:08 P.M.

At 4:06 P.M. the Sammis-Star 345-kV line tripped, initiating a large-scale cascading wave of line outages across the Northeast. A factor that caused the wave to spread was the extensive use of the transmission system to provide power from states south of Ohio to Ohio and Michigan. Regions in the south had surplus power available from relatively efficient generation plants. Demand in Detroit Edison's control area and First Energy's control area was significant. Rather than using inefficient units to provide power, Detroit Edison and First Energy would import less expensive power generated from the more efficient units in the south. A south-to-north power flow running through Ohio was common. On the day of the blackout, there were south-to-north flows. The power flows were all within system limits.

The wave was influenced by geography, and the geography created islands that were disconnected from the Eastern Interconnect. The first island was the city of Cleveland. With the sudden loss of imported power, the grid in the Cleveland area became unstable and generation units tripped off line, resulting in the blackout.

Transmission lines to the west of Cleveland were trapped in the cascading wave as they were unable to bear the burden of additional power flows. Eastern Michigan was cut off from Ohio. A line between bringing power to eastern Michigan from western Michigan also failed, and the only pathway into eastern Michigan was through Ontario to the northeast. Power from the south shifted to the east in an attempt to reach Michigan by circumventing Lake Erie in a counterclockwise path. This shifting of power caused a cascading failure of power lines between Pennsylvania and New York. The cascading wave pushed its way into New Jersey, completely cutting off New York from sources of power to the south.

The sole line into eastern Michigan from Ontario also tripped. Eastern Michigan, like Cleveland, became an island and blacked out for the same reasons that Cleveland did. Without incoming power from the south, power demand in New York City, looked for alternative sources. Power surged out of Ontario toward New York City, overwhelming the transmission lines. A series of lines tripped decoupled eastern New York from western New York. A similar situation occurred between eastern New York and the rest of New England. Eastern New York became an island that was disconnected from all neighboring systems. Eastern New York, including New York City, did not have the capacity to meet demand and blacked out just as Cleveland and Eastern Michigan had. Other areas that blacked out were the areas of northwestern Pennsylvania and Ontario. Western New York and all of New England with the exception of eastern New York managed to continue providing power service. From the onset of the Sammis-Star line trip, the cascading wave of transmission line trips lasted around 6 minutes, coming to an end at around 4:12 P.M.

Figure 3.9 shows the power flow patterns and blackout areas as the regional cascading wave evolved. The first diagram is the flow pattern just before the tripping of the Sammis-Star line. Around 510 generation units throughout the Northeast and Canada tripped off line as a result of the cascading wave.

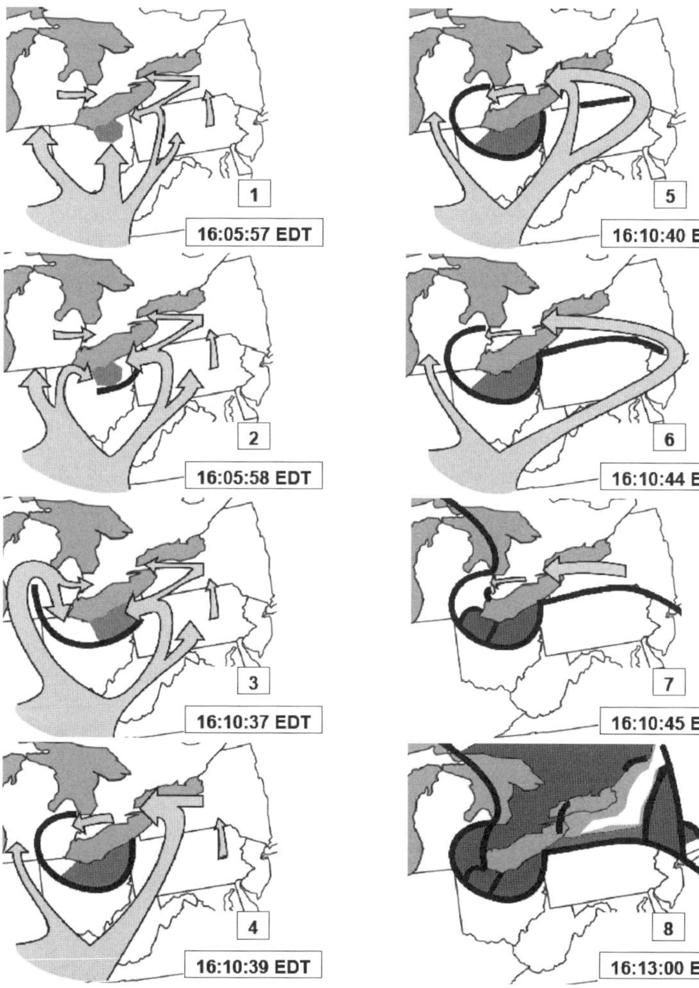

Figure 3.9

We conclude this section by placing the initial issues in the context of the blackout.

- This case illustrates the coordination of various agencies in order to maintain grid reliability (although such coordination was ineffective in this case).

The activities of a single control area within the Eastern Interconnect have widespread impact across the Interconnect. Power flows must be monitored and coordinated across control areas.

- This case shows the extent to which the transmission network is utilized.

Wholesale market activity in the Eastern Interconnect has increased exchanges across the Interconnect. It is not uncommon for power to flow from Ohio to Florida during winter months or from Florida to Ohio during summer months. The coordination among grid operators has become more critical as transmission use has increased.

- This case demonstrates how important it is to execute mundane operations properly.

The US-Canada Power System Outage Task Force jointly commissioned by the US and Canadian governments identified improper vegetation maintenance along transmission pathways as a factor causing the blackout. The cascading wave was initiated by the trip of four 350-kV lines, and three of these lines tripped because of line contact with trees. Proper pruning along the lines' pathways would have averted these line trips.

Additionally, appropriate servicing of system monitoring devices would have provided much needed information to MISO as well as First Energy's system operators. Because monitoring devices failed, operators were unaware of the state of the system and could not respond appropriately.

Chapter 4

Short-Term Utility Planning

This chapter provides the fundamentals of short-term planning and operations within a utility environment, assuming that each utility is responsible for operating the grid within its control area. Examining this in a utility setting is critical before moving to a market environment. The setting is simpler, allowing for a focus on operational requirements and objectives that any market design must also address.

The layout of this chapter is congruent with the layout of the previous chapters. In Section 4.1, we begin with a preliminary discussion of operations that frames the rest of the material. We then move on to address day-ahead demand forecasting in Section 4.2. Sections 4.3 through 4.7 address the day-ahead scheduling of units to meet demand requirements. We layer the complexity of unit commitment scheduling section by section. A simple example starts the discussion. Then realistic operational supply-side assumptions from the previous chapters are introduced incrementally. Short-term planning requires that two issues be addressed, cost and reliability. The cost issue is addressed first and is coupled with reliability as more realistic operational considerations are taken into account. Section 4.8 closes the chapter by addressing real-time operations and adjustments to the day-ahead schedule.

4.1 PLANNING AND EXECUTION OF DISPATCH: DAY-AHEAD PLANNING THROUGH REAL-TIME DELIVERY

This section provides the activities associated with power delivery from day ahead until delivery time. It provides the operational to do list for setting the dispatch and managing real-time load balancing. In doing so, the section frames the issues that are addressed in subsequent sections. Another important element of the section is Table 4.1. This table lists the operating parameters that characterize the plants available for dispatch.

Electric Power Planning for Regulated and Deregulated Markets. By Arthur Mazer

Table 4.1 Operating Parameters

Parameter	Definition
Minimum capacity	The minimum power output that a unit can operate at without causing unnecessary mechanical stress. The minimum capacity is measured in MW.
Maximum capacity	The maximum power in MW that the unit can provide
Optimal heat rate	The heat rate when a unit is operating at its optimal efficiency
Heat rate curve	The heat rate at various levels of power output
Incremental heat rate	The incremental heat rate provides the change in fuel requirement for a small increment in power output. The incremental heat rate is provided at various levels of power output to provide an incremental heat rate curve.
Ramp-up rate	The rate at which a unit can increase its power output once it has reached its minimum capacity
Ramp-down rate	The rate at which a generator can reduce power output. This may be different from the ramp-up rate.
Minimum uptime	The minimum time that a unit must be connected to the grid
Minimum downtime	The minimum amount of time that a unit must be in a rest state
Start up time	The time it takes a unit to connect to the grid from a rest state. Referring to the operating sequence in Section 4.3.2, the start-up time is the time that includes Steps 1 through 4. The start up time depends on the amount of time that the unit has been at rest. Longer rest times require longer start-up times because the boiler and tubes have cooled and require heating. Start-up times are frequently given for 3 regimes: hot start, cold start, and warm start. Definitions for these regimes must accompany the time.
Fuel start-up amount	The amount of fuel required to bring the unit from rest state to producing at minimum power. As with start-up time, this parameter depends on the length of time that the unit has been at rest and is given in 3 regimes: hot, warm, and cold start.
Start-up cost	The dollar amount that reflects the maintenance charge required for wear on the system due to a start-up
VOM	The variable operations and maintenance cost of a unit given in dollars per MWH of energy produced.
Forced outage rate	The percentage of time that a unit is not available because of equipment failure
Ramp-up sequence	The ramp-up sequence provides ramping rates and a heat rate curve for the operating regime below the minimum capacity. The operating characteristics in this regime differ from those above minimum capacity because the turbine passes through several zones of instability, causing vibrations, before attaining minimum capacity.

4.1.1 Day-Ahead Electricity Demand Forecasting

Day-ahead demand forecasting is the starting point for all other activities. Indeed, the purpose of all other activities is to supply the load. Without a forecast of the load no other activity can proceed.

Typically, a utility has a department that is responsible for day-ahead forecasting. This group maintains all the statistical factors that influence load. A day-ahead forecast requires quantifying all these factors and entering them into a model. The model produces a load forecast from the inputs. More detail follows in Section 4.2.

4.1.2 Day-Ahead Supply Forecast

Within a utility there are personnel responsible for generation, contract management of interutility contracts, contract management of fuel supply contracts, and dispatch planning. Dispatch planners must centralize up-to-date operational information of the utility's supply portfolio before setting a dispatch. Generation and contract management personnel pass on updated unit availabilities and outages, operating information, interutility contract costs, and fuel supply costs to planning personnel. Typically, utilities have information systems that allow for some level of automation of the information exchange. As mentioned above, there is a common set of parameters that describe operational information of generation units. Table 4.1 lists the parameters along with their definitions.

The information set may change daily. The main factors that cause changes in the supply portfolio include unit forced outages, scheduled maintenance of a unit, transmission outages, changes in interutility contract terms, fuel cost updates, weather impact on unit performance, and weather impact on transmission performance.

4.1.3 Day-Ahead Ancillary Services Forecasting

Dispatch planners must determine ancillary service requirements. As discussed in Chapter 3, ancillary service requirements depend on regulatory requirements as well as load and supply conditions. For example, a regulatory requirement might be that spinning reserve capacity must be sufficient to replace the largest operating generation plant in case this plant incurs operational difficulties and must be taken off line. This requirement varies in accordance with which units are available for dispatch.

4.1.4 Setting the Day-Ahead Dispatch Schedule

Once the above steps are complete, planners set an optimal dispatch schedule. This schedule is referred to as the least-cost dispatch or an economic dispatch. The least-cost dispatch is a scheduling of unit outputs that services load and satisfies ancillary service requirements with the lowest cost while also satisfying unit operating constraints as well as grid constraints. Determining the least-cost dispatch requires solving a problem known as a constrained optimization problem. These problems are easy to explain but difficult to solve. Solutions require millions of arithmetic operations that can only be performed by computer. Sophisticated computer algorithms are required to accurately solve problems in reasonable times. Every utility has in-house expertise in running these models and quality-checking the output. The models are referred to as dispatch simulators.

The experts must first ensure that the information from Steps 1, 2, and 3 is correctly input into the computer system that solves the least-cost dispatch problem. This information is stored in the dispatch simulator's database. As the information sets can be large, there are usually system protocols that allow for automated downloading between information sources and the dispatch simulation database.

Once the information has been centralized into the database, the planners prepare and execute the dispatch simulator to provide a least-cost solution. After execution of the model, the output is quality checked to ensure correctness.

4.1.5 Day-Ahead Scheduling of Utility-Owned Generation

Once a dispatch schedule has been set, planners forward it to day-ahead schedulers. There are generation schedulers, interutility contract schedulers, fuel schedulers, and transmission schedulers. The generation scheduler has the responsibility of informing each generation facility of its next day's operating schedule. Interutility contract schedulers inform their utility counterparty of required next-day power exchanges. Fuel schedulers inform their fuel suppliers of next-day fuel deliveries. Transmission schedulers schedule power flow exchanges across transmission lines of neighboring control areas.

4.1.6 Real-Time Adjustments to the Day-Ahead Schedule

Real-time adjustments occur for several reasons. These include load forecast errors, forced outages, transmission outages, and disruptions in power exchanges with neighboring utilities. Responses to these events are discussed in Chapter 3.

4.2 DAY-AHEAD DEMAND FORECASTING: LOAD AND ANCILLARY SERVICE REQUIREMENTS

Load forecasts are the starting point for the entire chain of power delivery. Planners require a load forecast before they can set an optimal dispatch. The load forecast must be of sufficient detail to provide necessary information to the dispatch model. For day-ahead scheduling hourly forecasts are necessary, although some planners use subhourly forecasts. An hourly forecast is established as an energy requirement in MWH or a power requirement in MW. For utilities with transmission congestion inside their control area, locational forecasts are also necessary. This section provides a step-by-step approach to developing a forecast and illustrates the approach with an example.

4.2.1 Approach to Developing Load Forecast

The following steps provide an approach to developing a load forecast.

Step 1. Identify Factors That Influence Load and Select the More Critical Factors as Variables

In Chapter 2 we identify factors that influence load. Variables are usually selected from this set.

Step 2. Select a Mathematical Model

There are two types of forecasting methods, top down and bottom up. Using statistical methods, each relates load to the load drivers presented in Chapter 2. Top-down forecasting aggregates all customers into a single load profile. A forecast model is set based on historical analysis of the aggregate load. Bottom-up forecasting disaggregates the customer base into different components: residential, industrial, and commercial. To apply bottom-up forecasting, meters are placed in a sample of the customer base; data are collected and archived in real time. The sample contains a representative cross section of residential, industrial, and commercial customers. Forecasters base their predictions on models that process this data.

In a utility environment, where aggregate information is available top-down forecasting is effective. There are many companies who sell load forecasting models commercially. Two types of models are most popular: regression models and neural networks. Regression models produce excellent results and have the advantage that they provide statistical information on errors. In the remainder of this section we provide a simple example of a load forecast and use this example to illustrate how models are created and improved.

Linear regression is the simplest form of regression. As an example, suppose we wish to predict the maximum load for the next day. The equation for load is the following:

$$L = a_0 + a_1X_1 + a_2X_2 + a_3X_3 + \ldots + a_nX_n$$

- L represents the maximum load.
- X_j are variable inputs, such as temperature, temperature change, humidity, dew point, and others.
- a_j are constants. These are determined by solving for the set of a_j that best match the historical data.

Step 3. Calculate Model Parameters, Determine Modeling Errors, and Decide Whether the Errors Need to Be Addressed

The solution of the a_j in the previous step is well known. Many software programs are available that compute the solution along with other relevant statistical information. The statistical information assists in determining how well the model fits the data. Additionally, the information informs the modeler of the significance of the variables; one can determine which variables are most significant and which are insignificant.

Step 4. If Necessary, Refine the Model and Return to Step 3; Otherwise, the Model is Complete

If the conclusion from Step 3 is that the model requires revision, there are several possibilities to consider.

- A single linear equation may not be appropriate for describing the behavior of power consumers. Several linear equations might make a better fit. For example, if X_1 is the peak temperature, a single linear equation clearly does not work. In the summertime, higher temperatures increase demand by increasing air conditioning usage. Accordingly, the corresponding coefficient, a_1, is positive. However, in the wintertime higher temperatures reduce demand because heating requirements are not as great. Accordingly, a_1 is negative. Clearly, one must develop separate equations for summertime and wintertime loads.
- The behavior of the load is not linear in its variables. In this case, nonlinear terms should be considered. For example, in the regression equation, set $X_2 = X_1^2$. There are different statistical methods that one can apply to determine whether or not to incorporate nonlinear terms. One very straightforward method is to plot the data against the variable of interest. If the data have apparent curves, then nonlinear terms may be considered.
- The variables that have been considered are not complete enough to describe the data. Other factors as described in Chapter 2 might need to be considered. These factors include additional weather variables as well as behavioral variables that account for demand changes due to social behavior, for example, weekend versus weekday consumption patterns.

The final model may incorporate a combination of all three of these revisions.

EXAMPLE *Peak Summertime Load*

In this example the above-described steps are applied to modeling the peak daily summertime load of a hypothetical control area. The data set can be viewed in the graph of Figure 4.1. The dots on the graph indicate load observations in MW along with associated daily maximum temperatures.

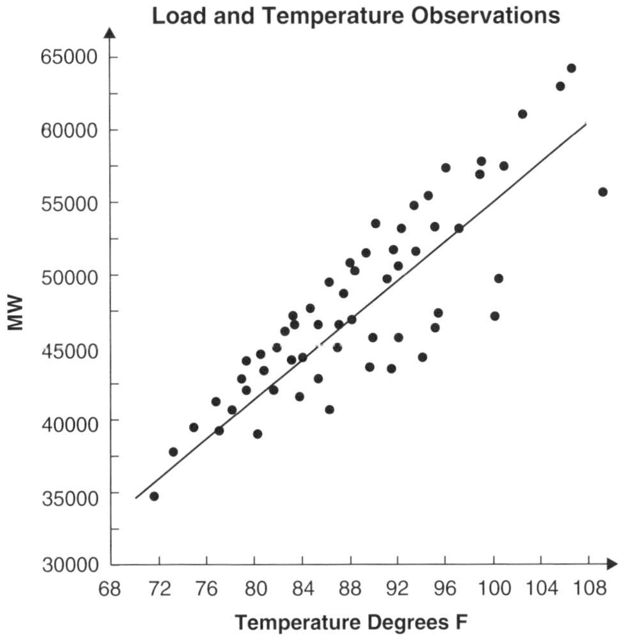

Figure 4.1

Step 1. Identify factors that influence load and select the more critical factors as variables.

We select a single variable, the daily maximum temperature.

Step 2. Select a model for the critical factors.

We assume that there is a fixed component to the load due to factors such as industrial machinery and lighting. Then there is a variable component due to temperature. Since the variable component of load is predominantly due to air conditioning demand, and air conditioning demand responds linearly with temperature, a linear regression is appropriate.

$$L = a_0 + a_1 X$$

The parameter a_1 represents the additional power requirement in MW for every degree Fahrenheit change. X is a variable that represents the daily high temperature.

Step 3. Calculate model parameters, determine modeling errors, and decide whether the errors need to be addressed.

The load and temperature data were loaded into a spreadsheet. The spreadsheet is linked to a program that is capable of calculating the parameters a_0 and a_1. The results are $a_0 = -1300$ and $a_1 = 680$. Figure 4.1 displays the regression line along with the data. Recall that the horizontal axis is the daily high temperature in degrees Fahrenheit and the vertical axis is the daily peak load in MW.

Step 4. If necessary, refine the model and return to Step 3. Otherwise, the model is complete.

Observing the differences between the regression line and the data, we note that the data fit well with the exception of some outliers. We wish to refine the model so that there are fewer outliers. There are several possibilities to consider. One could consider including higher-order terms. Since there are no apparent curves in the data, this does not look promising and a straight line approximation appears appropriate. Another consideration is to include more variables. More weather variables may be of interest. These would include overnight low temperatures, temperatures of the preceding day, humidity, and cloud coverage among others. A final consideration is to look at the outliers and see whether there is a recognizable pattern that might be explained by consumer behavior.

Following the final consideration, an investigation of the outliers reveals that many of the data points far below the regression line occur on holidays and weekends. This suggests that two regressions are required: one applying to weekdays only and the other for holidays and weekends. We carry this out for the weekday regression only. Eliminating the weekend and holiday data and recalculating the regression coefficients provides the following result: $a_0 = -1700$ and $a_1 = 750$. Figure 4.2 provides a plot of the resulting regression line along with the corresponding weekday data.

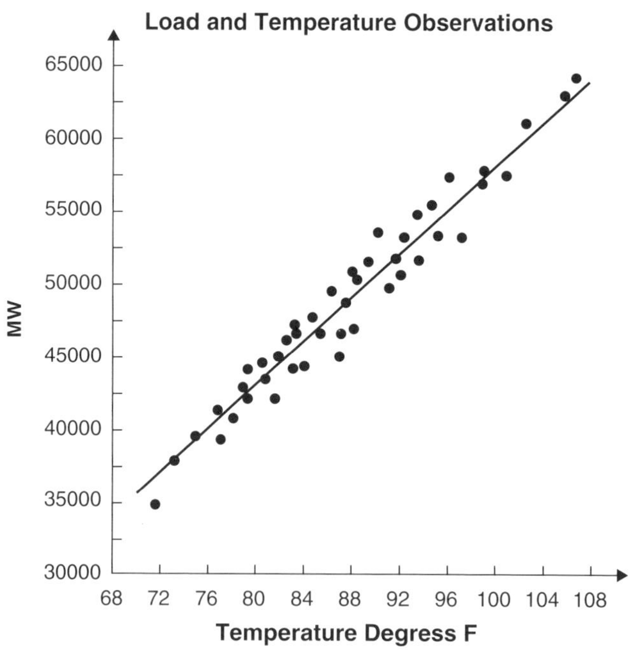

Figure 4.2

Visually, one can see that this is an improvement over the former result. If further improvements are desired, other approaches as listed above must be considered.

We close this section with some final remarks.

Remarks

- Ancillary service requirements must be added to load requirements. Reserve requirements are generally independent of load. However, regulation requirements are dependent on the load forecast.
- Experienced forecasters have developed excellent models that are very accurate. Even with these excellent models, forecasts may be a bit off the mark. The greatest source of error is error in the weather forecast. When making a forecast of next-day load, actual weather is not available; forecasted weather must be used. If the weather forecast is inaccurate, the load forecast will also be inaccurate. Many utilities hire weather consultants and subscribe to various weather services, hoping to find the best forecast.
- The model above did not include the hourly resolution that is required for daily processes. There is a great distance between a daily peak forecast and an hourly forecast. Hourly models require many more variables along with considerably more data to select the parameter values.
- An hourly forecast of energy requirements does not provide power requirements to meet continuously fluctuating power demands. The fluctuations are handled through load-following ancillary services. A demand forecast requires a forecast of ancillary service requirements in addition to the hourly load forecast.
- A complete demand forecast must also include a forecast of reactive power requirements. This forecast is placed into a model that tests dispatch plans to ensure that they respect grid constraints as discussed in Section 4.7.
- Data from the most recent loads provide the most current responses to the selected variables. However, there is not sufficient data from recent observations to perform a robust statistical analysis and accurately determine parameters. Accordingly, historical data from previous years must be incorporated into the data set. The choice of which data to select and how to incorporate them into the model affects results and is of concern.

4.3 LEAST-COST DISPATCH IN A SINGLE CONTROL AREA: A SIMPLE MODEL

This section provides a simple model for determining the day-ahead dispatch schedule. The model is not intended as an actual proposal for setting the dispatch. Instead, it is a starting point for understanding actual constructions. Behind every simplified model are simplifying assumptions. This section begins with a set of assumptions. Then the section concisely poses the optimization problem that provides the dispatch as a solution. Finally, the simplified solution is presented.

4.3.1 Assumptions

- Transmission charges and losses for all resources are negligible.
- There are no imports or exports between the control region and any of its neighbors.
- We ignore all network requirements, e.g., primary reserve, secondary reserve, voltage support, voltage regulation.
- The only operating constraint on each unit is the maximum capacity. The units may be started, stopped, ramped up, and ramped down at any time to any output level up to their maximum capacity.
- The heat rate of any unit is a fixed constant under all operating conditions.
- The fuel costs are determined by the heat rate and fuel prices. All other costs can be expressed as a constant variable operational and maintenance (VOM) cost in $ per MWH.
- An hourly forecast is sufficient to set the dispatch. Load fluctuations within a given hour are immaterial. The hourly forecast is set in MW as a constant power requirement.

4.3.2 Optimization Problem Definition

A well-defined optimization problem contains two features that are presented concisely as mathematical statements. These two features are the objective function and the constraints. The objective function provides the goal that one wishes to attain along with the decision variables that are available for attaining that goal. The constraints describe explicit and implicit restrictions on the decision variables. Experience shows that if one is unable to describe the problem in words it is unlikely that one will arrive at a concise mathematical description. Throughout this chapter, we first provide a textual description of optimization problems followed by a mathematical description.

In Words

- Objective

  The objective is to minimize the total cost of the dispatch. The total cost includes fuel costs and VOM charges. The decision variables available for attaining this objective are the MWH production of each unit by hour.
- Constraints

  1. The total power output must match the load forecast.
  2. Each unit has a maximum power output that it may not exceed.

Note what would happen if the constraints were not identified. The solution is to not dispatch anything and not incur any costs. But then load would not be satisfied. In life we are always under constraints when we wish to accomplish any objective;

the constraints are an integral part of the problem. The constraints here are very simple because of the assumptions. In later sections, the constraints are more complicated.

In Equations

- Objective

$$\min_{\text{output}_{jh}} \sum_{h=1}^{24} \sum_{j=1}^{N} \text{output}_{jh}(F_j + \text{VOM}_j)$$

- Constraints:

1. $\sum_{j=1}^{N} \text{output}_{jh} = L_h$ for each h

2. $0 \leq \text{output}_{jh} \leq M_j$ for each pair jh

An explanation of the equations is as follows:

- Min is a mathematical expression that stands for the operation "minimize."
- The expression under Min provides the decision variables that are available for accomplishing the minimization. In this case, output_{jh} is the output of unit j during hour h and the decision is over all units for every hour.
- The Greek symbol sigma represents a sum. Associated with a sum is an index that provides the scope of the sum. Below are examples that illustrate the notation.

1. $\sum_{j=1}^{3} j = 1 + 2 + 3 = 6$

2. $\text{output}_1 = 3$, $\text{output}_2 = 7$, $\text{output}_3 = 2$

$$\sum_{j=1}^{3} \text{output}_j = 3 + 7 + 2 = 12$$

3. $\text{output}_{11} = 3$, $\text{output}_{12} = 2$, $\text{output}_{21} = 5$, $\text{output}_{22} = 1$

$$\sum_{h=1}^{2} \sum_{j=1}^{2} \text{output}_{hj} = 3 + 2 + 5 + 1 = 11$$

- Note that the h index ranges from 1 to 24, indicating 24 hours in the day, and the j index ranges from 1 to N, indicating that there are N units available in the system.
- The expression $\text{output}_{jh}(F_j + \text{VOM}_j)$ is the cost of running unit j at output_{jh} during hour h. F_j is the fuel cost. This is the heat rate of unit j multiplied by the cost of the fuel. VOM_j provides the variable operational and maintenance charges in \$ per MWH.
- The first constraint set states that for any hour h, the sum of the outputs of each unit must equal that hour's load, L_h. The load must be satisfied.

- The second set of constraints states that for any unit j, and any hour h, the output must be between zero and the maximum capacity of the unit, M_j. Note that there are $24 \times N$ constraints in the second constraint set.

A Numerical Example

In the following paragraphs we provide a load forecast and a portfolio of units and determine the least-cost dispatch for the simplified problem.

Let the forecasted hourly loads for the next day be given by the graph of Figure 4.3.

The set of units available for generation is called the supply stack. The supply stack and unit characteristics for this example are provided in Table 4.2.

The total variable cost is a cost in \$ per MWH and is the sum of the fuel cost and the VOM cost. Note that the units are arranged from the least expensive total variable cost to the most expensive total variable cost. This is the order of selection for satisfying load and is known as merit order. It is common to plot the supply stack in order to show the total variable cost as a function of cumulative capacity. Such a plot is displayed in Figure 4.4.

A solution to the least-cost dispatch is available with the use of the graph. For each hour, tick off a point along the horizontal axis, cumulative capacity, that corresponds to the load requirement for that hour. All capacity to the left of this point is dispatched for that hour: all capacity to the right of that point is set to zero. No less capacity can be dispatched: otherwise, load would not be satisfied. If any other unit was dispatched it would be more costly. And so this solution solves the least-cost dispatch problem.

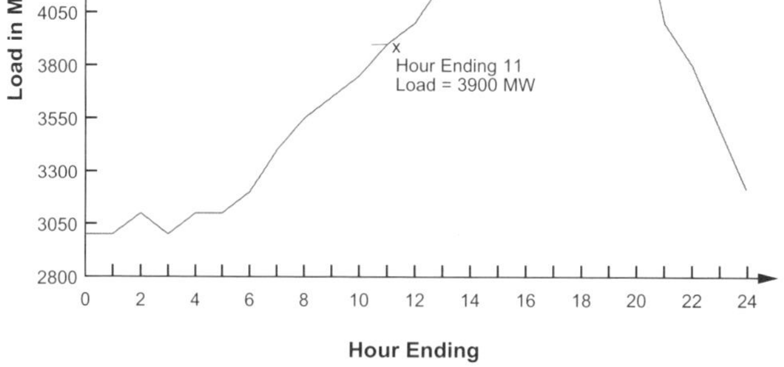

Figure 4.3

Table 4.2 Unit Characteristics

Unit name	Fuel	Max capacity	Heat rate	Fuel cost	VOM	Total variable cost
Aquarius	Hydro	320	NA	NA	2.25	2.25
Neptune	Nuclear	1200	10.1	1.50	6.00	21.15
Venus I	Coal	900	10.0	2.80	4.50	32.50
Venus II	Coal	800	10.3	2.80	4.50	33.34
Orion	Gas	550	6.9	7.45	4.00	55.41
Pegasus	Gas	500	7.0	7.60	4.00	57.20
Mercury	Gas	250	10.5	7.45	4.15	82.38
Mars	Gas	250	10.7	7.45	4.15	83.87
Pluto	Gas	150	11.3	7.45	4.15	88.34
Haley	Gas	150	12.5	7.45	5.00	98.13

NA, not applicable.

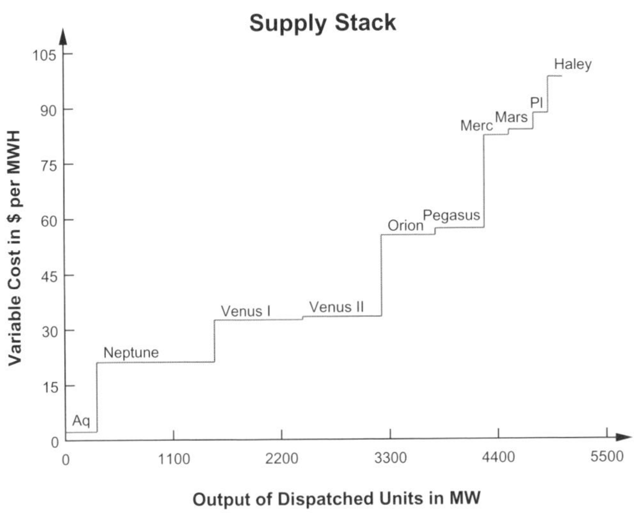

Figure 4.4

We provide an explicit expression of the solution for a single hour. For purposes of illustration examine the hour ending at 11. The energy requirement during this hour is 3900 MWH. Figure 4.5 graphically displays the solution.

The solution demonstrates that units Aquarius through Orion are fully dispatched, Pegasus is partially dispatched at the level that balances load, and Mercury, Mars, Pluto, and Haley are not dispatched.

A solution for the entire 24-hour dispatch can be displayed by stacking the units' capacities in merit order, cheapest to most expensive, and overlaying a plot of the load by hour. Figure 4.6 presents the solution.

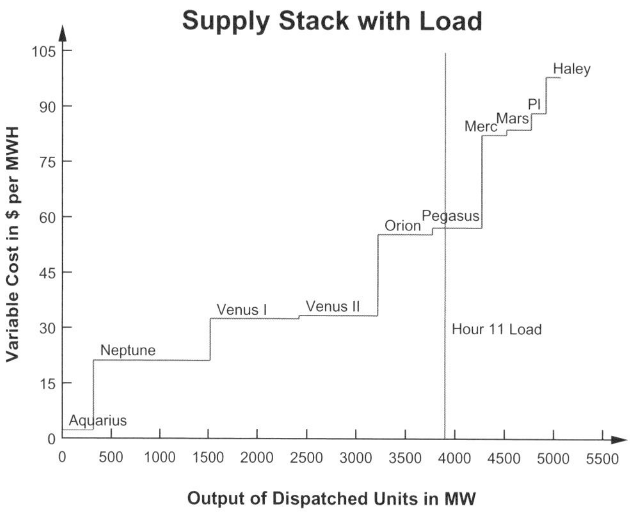

Figure 4.5

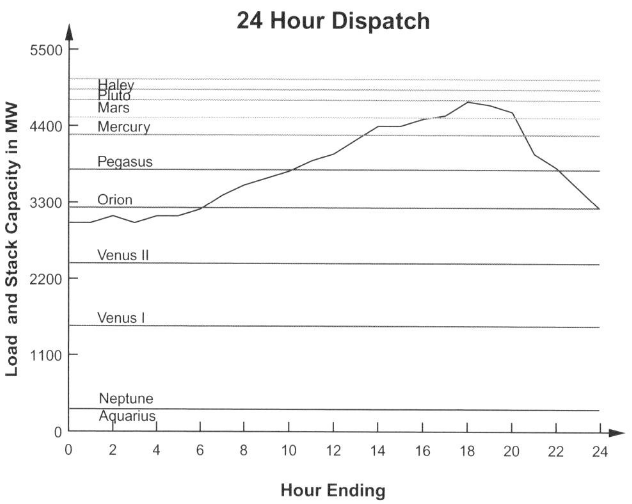

Figure 4.6

From the chart the dispatch schedule of every unit is available. Aquarius, Neptune, and Venus I are fully dispatched for the entire 24-hour period. Venus II is dispatched over the 24-hour period, but during the early morning it is not fully dispatched. Orion, Pegasus, Mercury, and Mars are dispatched during a portion of the day, while Pluto and Haley are never dispatched.

4.4 A SOLUTION USING PROFIT MAXIMIZATION

In this section we present another solution method for the least-cost dispatch problem. The assumptions from Section 4.3 apply. To illustrate the method we also use the example from Section 4.3. There are two motivations for presenting this method. The first motivation is that the method illustrates an actual approach for more complicated problems. The second motivation is that the solution illustrates a mechanism for balancing load and supply in competitive markets. We briefly remark on the latter issue; more details of market mechanics are addressed in Chapter 8.

The section begins with an important concept, system lambda. It then demonstrates the relation between system lambda and load. The relation suggests a related problem, the profit maximization problem. We introduce this problem and describe how to use it to solve the original least-cost dispatch problem.

Before proceeding we present one philosophical remark that explains the approach. Every profession has its common wisdom. In the modeling community, the wisdom is:

> *When there is a complex problem and if it is possible, break the problem down into a set of simpler problems that can be solved independently. Solve the set of simpler problems and build the entire solution as a composite of the individual solutions.*

This section's approach adopts the above common wisdom.

4.4.1 System Lambda

The system lambda is the marginal cost of production in \$ per MWH. Economists have strict definitions for marginal costs. In our case, the marginal cost refers to the cost in \$ per MWH of the most expensive unit that is dispatched in a least-cost dispatch. For the example in the previous section, the system lambda during hour 11 is the variable cost of the Pegasus unit, \$57.20 per MWH.

The relationship between the system lambda and the least-cost dispatch is that all resources that operate at a cost less than the system lambda are dispatched and all units that operate at a cost greater than the system lambda are off line. A unit with a cost equal to the system lambda is dispatched to the level required to meet load.

Suppose that a system lambda is known. With the above relationship, the least-cost dispatch can be set. In the example from the previous section, if we start with the system lambda for hour 11 at \$57.20 per MWH, we know to dispatch all units with a production cost of less than \$57.20 per MWH. We then know to balance the remaining load with Pegasus.

4.4.2 The Profit Maximization Problem

The above relationship suggests a problem related to the original least-cost problem. This problem is known as the profit maximization problem, and it is used to solve the original least-cost dispatch problem.

In Words

- Objective

  For a given hourly price series, maximize the profit of a given unit by controlling the unit's energy production.
- Constraint

  At any moment, the unit's power output may be no greater than the unit's maximum capacity.

In Equations

- Objective

$$\max_{\text{output}_h} \sum_{h=1}^{24} (\lambda_h - F - \text{VOM})\text{output}_h$$

- Constraint

$$0 \le \text{output}_h \le M \quad \text{for all } h$$

There are several points to note.

- The problem is defined for a single unit. This optimization problem is solved independently over all units to construct the solution for the entire portfolio of units.
- The relationship between the system lambda and the least-cost dispatch implies that all units independently solve the profit maximization problem against the system lambda. More details are presented below.
- In this problem, the system lambda is a set of 24 prices.
- For the simple problem, the constraints and the objective function act independently over every hour. This allows us to solve the problem over each hour independently. In subsequent sections, the operating assumptions are more realistic. They reflect features that impact the temporal operations of a unit, such as ramp-up rate and variable heat rates; the complete problem is not solvable by looking at each hour independently.

The first and second bullet points provide the key to the approach of this section. The method is iterative, and the steps are presented below.

Solution Algorithm

Step 1. Determine an initial estimate of the system lambda.

Step 2. Solve the profit maximization problem for each unit, using the estimated system lambda.

Step 3. Calculate the summed total output of all units and compare this against the load requirement.

Step 4. If the comparison of Step 3 is favorable, accept the dispatch from Step 2 as the least-cost solution. Otherwise, use the comparison of Step 3 to determine a new estimate of system lambda and return to Step 2.

There are several points to remark upon.

Remarks

- The strength of the method is that it allows one to solve the maximization problem independently for each unit as opposed to solving the entire problem for each unit and each constraint. This follows the common wisdom that we alluded to at the beginning of the section.
- There are numerous ways for accomplishing Step 1. For example, an experienced planner can present a reasonable estimate based on intuition. Alternatively, one can use the previous day's actual dispatch costs as an initial point.
- There are different ways to update the estimate. To do so requires a quantitative estimate of the relation between changes in output and changes in the system lambda. The better one can quantify this, the quicker one can converge to a solution.
- In practice one never arrives at a solution that perfectly matches load. Instead, the comparison in Step 4 is usually done with respect to a tolerance band around the load. One accepts a solution provided that the output lies within the tolerance band.
- One excellent question is whether or not the algorithm does indeed converge to a solution. Unfortunately, this is not the case. One reason that the solution may not converge is that the dispatch may be discontinuous in the system lambda: A small change in the system lambda leads to a large change in the total system dispatch. Consider, for example, the case in which a system has many peaking units with similar operating costs. As the system lambda increases above the cost of these similar units, they all dispatch. It may not be possible to balance supply with load by using the system lambda alone when the load requires some, but not all, of the peaking units. Nevertheless, in a regulated utility environment the algorithm provides a practical result that benchmarks well against other solution methods and can be appropriately adjusted by a dispatcher. The problems with convergence, however, do point to difficulties with competitive market mechanisms.
- Another interesting phenomenon that may occur is the nonuniqueness of the solution. Indeed, for the more complex problems posed in subsequent sections, there may be more than one set of system lambdas that produce the same dispatch. This is not of consequence to the day-ahead dispatcher: one still arrives at a least-cost dispatch. However, it is problematic when using the system lambda to develop market forecasts.

EXAMPLE ***The Simple Problem of Section 4.3***

We now apply the algorithm to the simple problem of the preceding section. Recall that the profit maximization problem can be solved by addressing each hour independently.

Accordingly, we only demonstrate a solution for a single hour's dispatch. The remaining hours are addressed identically. As in the previous section we select hour 11, in which $L_{11} = 3900$. The solution steps are as follows.

Step 1. Determine an initial estimate of the system lambda.

We start with a neutral guess, the production costs of the fifth plant, Orion. Set lambda = 55.41.

Step 2. Solve the profit maximization problem for each unit, using the estimated system lambda.

The solution is that all units with production cost lower than 55.41 dispatch at their maximum dispatch level while all units with production cost higher than 55.41 are idle. The unit with the same production cost as the system lambda, Orion, is indifferent to the level of dispatch; any output solves the profit maximization problem.

Step 3. Calculate the summed total output of all units and compare this against the load requirement.

The output can range between 3220 and 3770 MWH, depending on the dispatch of Orion. As the requirement is 3900, the output is insufficient.

Step 4. If the comparison of Step 3 is favorable, then stop; the dispatch from Step 2 is accepted as the least-cost solution. Otherwise, use the comparison of Step 3 to determine a new estimate of system lambda and return to Step 2.

Recognizing that output increases with increasing lambda, we will increase the system lambda to that of the 6th plant, Pegasus. Set the system lambda at lambda = 57.20.

Step 2. Solve the profit maximization problem for each unit, using the estimated system lambda.

The solution is that all units with production cost lower than 57.20 dispatch at their maximum dispatch level while all units with production cost higher than 57.20 are idle. The unit with the same production cost as the system lambda, Pegasus, is indifferent to the level of dispatch; any output solves the profit maximization problem.

Step 3. Calculate the summed total output of all units and compare this against the load requirement.

The output can range between 3770 and 4270 MWH, depending on the dispatch of Pegasus. As the requirement is 3900, setting the output of Pegasus at 130 MWH satisfies the requirement.

Step 4. If the comparison of Step 3 is favorable, accept the dispatch from Step 2 as the least-cost solution. Otherwise, use the comparison of Step 3 to determine a new estimate of system lambda and return to Step 2.

The dispatch is accepted and we are finished.

This example illustrates several points.

- A good initial estimate is helpful in arriving at a solution quickly.
- Another important factor for arriving at a solution quickly is the update of the estimate in Step 4. In this case, there is a very straightforward under-

standing of the relationship between system and the resulting dispatch lambda so that the update is effective.

- Because of the simplicity of the problem, the load is perfectly matched. This is not usually the case.

4.4.3 Final Comments on System Lambda and Markets

The introduction of markets into the power industry relies heavily on the profit maximization problem and the iterative algorithm for success. The idea is that traders for supply and traders for demand would bargain with one another in a market process that mimics the solution algorithm. If the outcome is a market price higher than the system lambda requires to satisfy demand, more units than are required to meet load would dispatch. There would not be enough purchasers on the buying side, and competition would cause the suppliers to move the market price toward the system lambda. Alternatively, if the market price is lower than the system lambda required to satisfy demand, an insufficient number of units to meet load would dispatch. Load purchasers would then increase their willingness to pay to the system lambda. Accordingly, the system lambda is the market price that is ultimately reached. The process does not function well in noncompetitive markets where producers can exercise market power; under noncompetitive conditions prices rise well beyond the system lambda. This issue is further addressed in Chapters 8 and 9.

4.5 LEAST-COST DISPATCH IN A SINGLE CONTROL AREA WITH OPERATING CONSTRAINTS

In this section, we expand on the least-cost dispatch model by incorporating operating constraints and more realistic cost functions. The format follows somewhat that of Sections 4.3 and 4.4. We start by presenting the assumptions and then write out the model formalization in words. Afterwards the model equations are presented. Finally, there are examples demonstrating the impact of operational characteristics on least-cost dispatch by technology type.

Several of the assumptions in Sections 4.3, and 4.4 are relaxed to provide an improved model. The first improvement is that production costs are more realistic. The cost function now varies with output. Additionally, start-up and shutdown costs are considered. The remaining improvements address operating characteristics. There is a more realistic view of the output by considering the minimum level of production. For hydro units, a total energy limit is considered. Finally, we incorporate characteristics that affect the temporal dynamics of the units. These include ramp-up and ramp-down rates as well as minimum uptime and minimum downtime. One point to note is that by relaxing the assumptions we increase the number of constraints required to more accurately model unit operating characteristics.

One evident impact of incorporating temporal operational characteristics is that the time horizon of the model is extended to 48 hours as opposed to 24 hours. This is not to develop a 48-hour dispatch plan, but to determine the off-peak dispatch of the units on day 1. The day 2 on-peak operating requirements impact day 1 off-peak decisions. Accordingly, a view of day 2 operations is necessary to set the correct off-peak day 1 least-cost dispatch. The effect of following-day dispatch requirements on next-day nighttime operations is demonstrated in the examples.

4.5.1 Assumptions

- Transmission charges and losses for all resources are negligible.
- There are no imports or exports between the control region and any of its neighbors.
- We ignore all network requirements, e.g., primary reserves, secondary reserves, voltage support, voltage regulation.
- An hourly forecast is used to set the dispatch. The power consumption within each hour is constant.
- The production of a unit below its minimum power threshold is negligible and considered to be zero.
- The fuel costs required to bring a unit up to minimum power are all contained in the start-up cost, while the fuel costs accrued during ramp-down from the minimum power threshold to zero are included in a shutdown cost.

4.5.2 In Words

- Objective

  The objective is to minimize the total cost of the dispatch. The total cost includes fuel costs, start-up costs, shutdown costs, and VOM charges. The decision variables available for attaining this objective are the energy production of each unit by hour.
- Constraints
  1. The total power output must match the load forecast.
  2. Each unit either provides no output or provides output bounded by minimum and maximum MW values.
  3. Each unit has a maximum ramp-up rate.
  4. Each unit has a maximum ramp-down rate.
  5. Once a unit is down, the unit has a minimum turnaround time.
  6. Once a unit is on line, the unit has a minimum on time.
  7. Units accrue a start-up cost for each start-up and a shutdown cost for each shutdown.
  8. The hydro facility has a maximum total energy availability.

4.5.3 In Equations

- Objective

$$\underset{\text{output}_{jh}}{\text{Min}} \sum_{h=1}^{48}\sum_{j=1}^{U} \text{cost}_j(\text{output}_{jh}) + \text{u}_{j,48}\text{SU}_j + \text{d}_{j,48}\text{SD}_j$$

  - $\text{cost}_j(\text{output}_{jh})$ is the cost incurred by unit j during hour h while producing output_{jh} MWH.
  - SU_j is the cost of a single start-up for unit j.
  - SD_j is the cost of a single shutdown for unit j.
  - $u_{j,48}$ is the total number of start-ups that occur over the time horizon.
  - $d_{j,48}$ is the total number of shutdowns that occur over the time horizon.
- Constraints

1. $\sum_{j=1}^{U} \text{output}_{jh} = L_h$ for each h

2. $\text{output}_{jh} = 0$ or $m_j \le \text{output}_{jh} \le M_j$ for each pair jh

3. If $\text{output}_{jh} - \text{output}_{j,h-1} > 0$ then $\text{output}_{jh} - \text{output}_{j,h-1} < \text{ru}_j$
  - ru_j is the maximum ramp-up rate for unit j.

4. If $\text{output}_{j,h-1} - \text{output}_{jh} > 0$ then $\text{output}_{j,h-1} - \text{output}_{jh} < \text{rd}_j$
  - rd_j is the maximum ramp-down rate for unit j.

5. If $\text{output}_{jh} > 0$ and $\text{output}_{j,h+1} = 0$, then $\text{output}_{j,h+k} = 0$ for all $k \le \text{dt}_j$
  - dt_j is the minimum downtime for unit j.

6. If $\text{output}_{jh} = 0$ and $\text{output}_{j,h+1} > 0$, then $\text{output}_{j,h+k} > 0$ for all $k \le \text{ut}_j$
  - ut_j is the minimum uptime for unit j.

7.

  a. $u_{j0} = 0$. If $\text{output}_{j,h-1} = 0$ and $\text{output}_{jh} > 0$ then $u_{jh} = u_{j,h-1} + 1$; otherwise, $u_{jh} = u_{jh-1}$

  b. $d_{j0} = 0$. If $\text{output}_{jh} = 0$ and $\text{output}_{j,h-1} > 0$ then $d_{jh} = d_{j,h-1} + 1$; otherwise, $d_{jh} = d_{jh-1}$
  - u_{jh} is the number of start-ups accrued by unit j up to hour h.
  - d_{jh} is the number of shutdowns accrued by unit j up to hour h.

8. $\sum_{h=1}^{48} \text{output}_{kh} \le w_k$ for all hydro units.
  - w_k is the total energy availability of the kth hydro facility.

Remarks

- In the objective function the unit cost, $\text{cost}_j(\text{output}_{jh})$, is a function of output. Cost functions are frequently modeled as cubic polynomials. Optimal coefficients for the terms in the cubic polynomial are selected to match historical data over the output range.
- Typically, the number of start-ups and shutdowns in 1 day doesn't exceed 1, but there are exceptions.

4.5.4 Examples

The following examples provide typical dispatch outcomes for specific technologies. The examples move through the supply stack in m erit order.

EXAMPLE 1. *Hydroelectric Power and Peak Shaving*

We consider the least-cost dispatch of hydroelectric power under energy-constrained circumstances. This is constraint number 8 in the equations. The constraint limits the amount of water that flows through the turbines; this reflects the need to maintain reservoir levels. Because hydro facilities operate at similar efficiency for all outputs, there is no efficiency gain at any specified output. Frequently the economic dispatch of hydro facilities follows a process known as peak shaving. Peak shaving is the utilization of hydro power during peak loads to flatten the load portion that is not served by the hydro facility.

Peak shaving can be used to avoid the start-up cost and fuel cost of an expensive peaking unit. We illustrate the concept using the example from the previous section while setting the maximum energy production of the hydro facility at 900 MWH. For ease of illustration we do not incorporate additional complexities of the other units in the example; the only additional constraint is the maximum energy constraint on the hydro facility. For completeness, the load and costs of the units are once again presented.

Figure 4.7 provides a graph in which all the units except the hydro facility are stacked against the load in merit order. From the graph it is seen that all the units including Pluto and Haley are dispatched to satisfy load.

The graph demonstrates that without hydro production the most expensive unit to come on line is Haley. The power requirement to displace Haley is approximately 150 MW, and the energy requirement is 250 MWH. This is the energy production from Haley during hours ending at 18 and 19. Hydro can completely displace this unit, saving a start-up cost and fuel costs. The second most expensive unit is Pluto. The level of power required to displace both

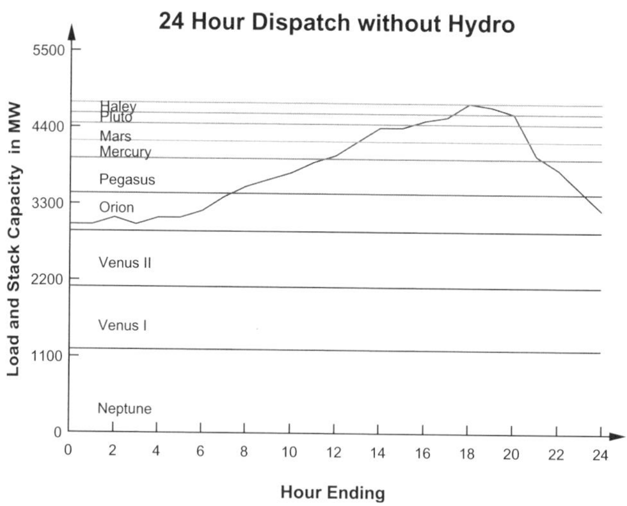

Figure 4.7

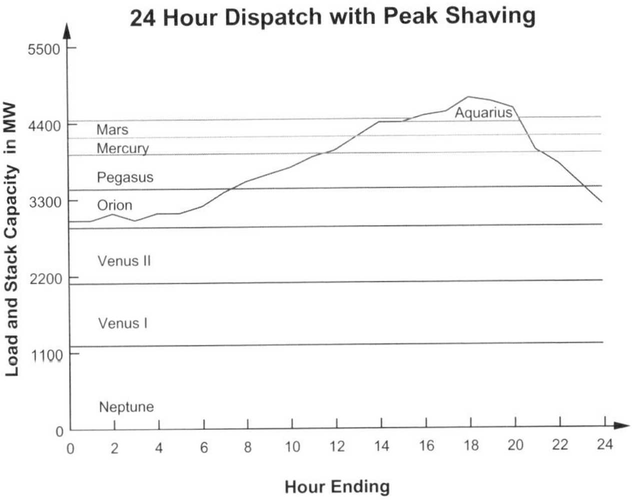

Figure 4.8

Haley and Pluto is 300 MW. The volume of energy required to displace both is around 850 MWH. This represents all the energy production from the Pluto and Haley units from the hour ending at 16 through the hour ending at 20. The hydro facility has the capacity as well as sufficient energy availability to displace both of these units.

After displacing both Pluto and Haley, the hydro facility has a surplus energy of 50 MWH. These 50 MWH are most optimally utilized by displacing 50 MWH of the next most expensive resource, Mars. We do this while maintaining the constraint of Aquarius' power output, 320 MW. In Figure 4.8, the area labeled Aquarius represents the energy supplied by the hydro unit. This total area is 900 MWH, representing the maximum energy that Aquarius can produce. The maximum distance between the load line and the base line for Aquarius is less than 320 MW, demonstrating that the power constraint of Aquarius is satisfied. Note that Pluto and Haley are not included in the dispatch.

EXAMPLE 2. *Nuclear*

In the US, nuclear units typically operate at constant capacity around the clock. The fuel cost associated with the units is most favorable, and system load is lower than the output capacity of the nuclear units. Accordingly, there are no nighttime load constraints. Although fuel costs are inexpensive, maintenance costs for nuclear units are very expensive. This motivates running the units in a least stressful manner: constant output. In France, 78% of electric production is from nuclear units. The French own specially designed reactors that allow them to cycle with load.

EXAMPLE 3. *Baseload Coal and Optimization Using Marginal Costs*

During the daytime, most baseload coal units operate at their optimal power level. During the night, load requirements are often insufficient to require that all baseload units operate at optimal capacity levels. However, baseload coal units are not very flexible; once they are off line it normally takes quite a long time to bring these units back on line. Additionally, start-up costs are significant. For these reasons baseload coal units are left operating throughout the off-peak period so that they are available to meet on-peak demands. There are several possible outcomes for optimizing nighttime operations. This example illustrates some outcomes. The example is more mathematically oriented than others and may be skipped without affecting the reader's understanding of subsequent material.

Assume unit characteristics as provided in Table 4.3.

In this example, we calculate optimal nighttime operations for the coal plants under three cases. For each case the nuclear plant is kept running at its constant output, 1200 MW. We note that for both coal units the minimum downtime is 7 hours, and we assume that this is insufficient time to take either coal unit off line, bring it back on line, and ramp it up to economically satisfy the next day's on-peak requirements.

To calculate the optimal coal plant outputs it is necessary to specify the cost curve and marginal cost curves of the units. The cost curve is designated by $C(X)$ where X is a variable in MWH. It provides a dollar cost for generating X MWH of energy with a constant output over an hour's time. This includes fuel and VOM charges; however, it does not include a start-up cost.

The marginal cost curve is represented by $MC(Y)$ and defined by the following relationship.

$$C(X + Y) = C(X) + MC(X) \times Y \qquad \text{whenever } Y \text{ is small.}$$

The units of the marginal cost curve are dollars per MWH. Both X and Y are quantities of energy in MWH.

For those familiar with calculus, the marginal cost curve is the derivative of the cost curve. In our context, Y is small: $Y = 1$ MWH. An interpretation of the marginal cost curve is that it is the cost of producing one additional MWH of power beyond X MWH. Components that make up the costs of the marginal cost curve include fuel and VOM. Fuel is the predominant component. The fuel component of the marginal cost curve is the incremental heat rate multiplied by the fuel price.

In this example we assume the following cost curves for Venus I, $C_1(X)$, and Venus II, $C_2(X)$:

$$C_1(X) = .000001X^3 - .0033X^2 + 40X + 10{,}000 \qquad 450 \leq X \leq 1200$$
$$C_2(X) = .0000004X^3 - .0012X^2 + 38X + 8000 \qquad 400 \leq X \leq 1100$$

Figure 4.9 displays the graphs of the cost curves for Venus I and Venus II.

The marginal cost curves are as follows:

$$MC_1(X) = .000003X^2 - .0066X + 40 \qquad 450 \leq X \leq 1200$$
$$MC_2(X) = .0000012X^2 - .0024X + 38 \qquad 400 \leq X \leq 1100$$

Table 4.3 Supply Table

Unit	Fuel	Min capacity	Max capacity	Ramp-Up rate	Minimum downtime
Neptune	Nuclear	700	1200	200 MW/h	48 h
Venus I	Coal	400	1200	250 MW/h	7 h
Venus II	Coal	450	1100	250 MW/h	7 h

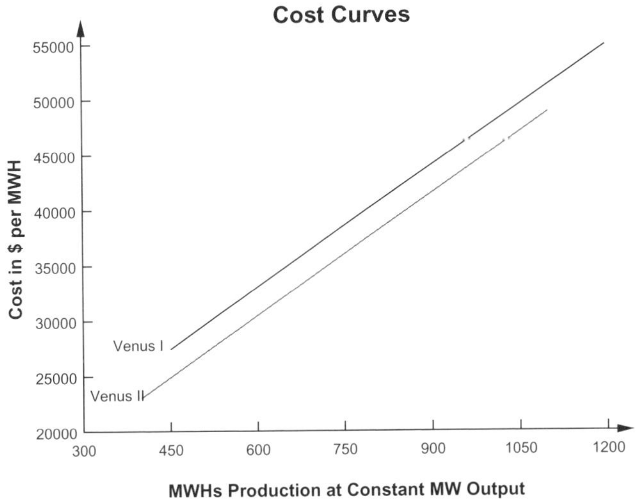

Figure 4.9

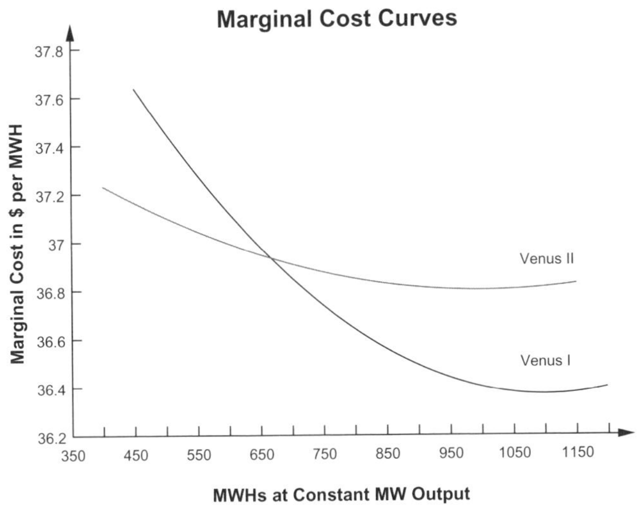

Figure 4.10

The graph of the marginal cost curves is provided in Figure 4.10.

We demonstrate the relationship between the cost and marginal cost curve, using Venus I's curves at X = 1000 MWH. Using the cost curve to calculate the cost of producing 1001 MWH gives the result that C(1001) = \$47,736. Using the cost curve and the marginal cost curve to calculate the cost of producing 1001 MWH gives the following result:

$$C(1000) + MC(1000) \times 1\,\text{MWH} = \$47{,}700 + \$36 = \$47{,}736$$

The two results are the same, and the cost of producing the additional MWH is \$36.

It is possible to construct an iterative solution to the optimization problem as follows.

Step 1. Select an initial production level for Venus I and Venus II that satisfies the total production constraint.

Step 2. Calculate the marginal costs of the selection.

Step 3. If possible, increase the production of the unit with cheaper marginal cost by 1 MWH and decrease the production of the unit with more expensive costs by the same amount and return to Step 2. Otherwise, a candidate solution has been identified.

There are some points to note.

Remarks

- Step 3 allows one to reduce costs by incrementally shifting production from the higher-cost marginal MWH to the lower-cost marginal MWH.
- There are two ways to reach a solution. First, one might run into an operating constraint; one of the units may be at its minimum (maximum) output and the output may not be decreased (increased). Second, the two units may have the same marginal cost, so there is no benefit in shifting production of the marginal MWH.
- The above point illustrates that there are two scenarios for candidate solutions. The first is that one of the plants operates an operating limit while the other satisfies the production constraint. The second scenario is that both units run at the same marginal cost.
- Solutions are identified as candidate solutions for technical reasons. The candidate solutions may not be unique and must be compared to other candidate solutions. For this example, the candidate solutions are actual solutions.

We illustrate this solution method for three different cases. The different cases demonstrate different solution scenarios.

Case 1. Load = 2200 MWH; 1000 MWH are required from the coal units.

Step 1. Select an initial production level for Venus I and Venus II that satisfies the total production constraint.

Initially, set the production of both units at 500 MWH to satisfy the 1000-MWH need.

Step 2. Calculate the marginal costs of the selection.

$$MC_1(500) = \$37.45/MWH$$
$$MC_2(500) = \$37.10/MWH$$

Step 3. If possible, increase the production of the unit with cheaper marginal cost by 1 MWH and decrease the production of the unit with more expensive costs by the same amount and return to Step 2. Otherwise, a candidate solution has been identified.

Following this step, one decreases the production of Venus I and increases the production of Venus II.

An examination of the marginal cost curve indicates that iterating through Steps 2 and 3 of the process continually brings about the result that Venus I's production is decreased while Venus II's production is increased. This occurs until Venus I's minimum output operating constraint is attained and it is not possible to reduce Venus I's production any further. The solution is as follows: Venus I produces 450 MWH, while Venus II produces 550 MWH.

Case 2. Load = 3200 MWH; 2000 MWH are required from the coal units.

Step 1. Select an initial production level for Venus I and Venus II that satisfies the total production constraint.

Initially, set the production of both units at 1000 MWH to satisfy the 2000 MWH need.

Step 2. Calculate the marginal costs of the selection.

$$MC_1(900) = \$36.40/\text{MWH}$$
$$MC_2(900) = \$36.80/\text{MWH}$$

Step 3. If possible, increase the production of the unit with cheaper marginal cost by 1 MWH and decrease the production of the unit with more expensive costs by the same amount and return to Step 2. Otherwise, a candidate solution has been identified.

In this case we increase the production of Venus I and decrease the production of Venus II. Continuing the process leads to the result that Venus I operates at its maximum output, 1200 MW, and Venus II operates at 800 MW.

Case 3. Load = 3200 MWH with different cost curves; 2000 are MWH required from the coal units.

In this problem we consider the following cost and marginal cost functions.

$$C_1(X) = .000001667X^3 - .0045X^2 + 40X + 10{,}000 \qquad 450 \le X \le 1200$$
$$C_2(X) = .000000444X^3 - .0012X^2 + 37X + 8000 \qquad 400 \le X \le 1100$$

$$MC_1(X) = .000005X^2 - .009X + 40 \qquad 450 \le X \le 1200$$
$$MC_2(X) = .00000133X^2 - .0024X + 37 \qquad 400 \le X \le 1100$$

MC_2 is different from the previous cases. Figure 4.11 provides graphs of this pair of marginal cost curves.

We next apply the solution steps to this case.

Step 1. Select an initial production level for Venus I and Venus II that satisfies the total production constraint.

Initially, set the production of both units at 1000 MWH.

Step 2. Calculate the marginal costs of the selection.

$$MC_1(1000) = \$36.00/\text{MWH}$$
$$MC_2(1000) = \$35.93/\text{MWH}$$

Step 3. If possible, increase the production of the unit with cheaper marginal cost by 1 MWH and decrease the production of the unit with more expensive costs by the same amount and return to Step 2. Otherwise, a candidate solution has been identified.

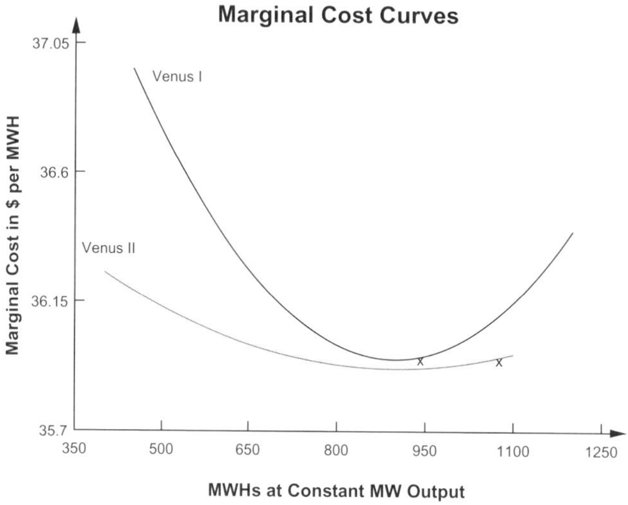

Figure 4.11

Following this step, one decreases the production of Venus I while increasing the production of Venus II.

Iterating Steps 2 and 3, one arrives at the solution to dispatch Venus I at 935 MW over the hour and dispatch Venus II at 1065 MW over the hour. Note from the marginal cost curves that for these output levels, the marginal costs of each unit are identical at $35.95/MWH. These points are marked on the marginal cost graph with an x.

We close this example with some remarks.

Remarks

- Although the example illustrates the solution across two units, the result applies when there are many units as well. In a dispatch involving many units, the most efficient dispatch at their maximum capacity. The marginal units, those that operate close to the system lambda, have solutions as demonstrated in this example. Either they operate at some physical limitation, minimum or maximum capacity, or they operate at the same marginal costs.
- An interesting phenomenon to note from the example is the property that the unit with maximal output switches depending on requirements. In Case 1 Venus II provides more output than Venus I. However, in Case 2 Venus I provides more output than Venus II. This role reversal comes about because the marginal cost curves of the two units intersect with one another. At lower output values, Venus I's marginal cost is lower than Venus II's, whereas at higher output values the opposite holds. This is not uncommon in actual units.
- The marginal cost curve is not independent of the cost curve. Indeed, one derives the marginal cost curve from the cost curve.

EXAMPLE 4. *CCGTs, Baseload or Cycling*

CCGTs are generally more flexible than baseload coal but less flexible than CTs. They are dispatched during the on-peak time frame based on their merit order. When gas prices were much lower than the current prices, CCGTs were economical as baseload units; they were on line around the clock. In most parts of the US, current gas prices do not merit CCGTs as baseload units, although in some countries such as The Netherlands there are no alternatives; some CCGTs operate as baseload units to meet demand. The decision of whether or not to run them as baseload units is a trade-off between the operating cost to run them overnight and the full start-up cost of the next day. This example examines the decision.

We consider the following assumptions.

- Aside from the CCGT, there is a nuclear unit and two coal units, all having the same characteristics as those of the previous examples.
- The CCGT has a two-on-one configuration: two CTs on one steam turbine. The CCGT has a minimum operating level of 150 MW. At the minimum operating level the heat rate is 11 and the VOM charge is negligible. The configuration for the minimum operating output is one CT on one steam turbine.
- A complete start-up for the CCGT is \$25,000, and a start-up for a single CT is \$10,000.
- Load requirements are such that a CCGT is not needed after 11 P.M., but ramp-up constraints require that all three generators of the CCGT are on line by 6 A.M. in order to service the next day's on-peak load.
- The off-peak load requirement is 3200 MW across all hours.
- The nuclear unit runs at a constant output, just as in the previous example. Accordingly, there are 2000 MW of demand to optimize between the coal units and the CCGT.
- The economic dispatch of the next day from 6 A.M. onward is identical for the options we consider. Accordingly, the only dispatch differences arise between 11 P.M. and 6 A.M.
- The cost curves of the coal units are the same as Cases 1 and 2 in the previous example.

We determine the optimal off-peak CCGT and coal unit dispatches from among the following two options.

Option 1. Take the CCGT off line during the off peak until 6 A.M. next day. Assume a start-up charge of \$25,000 for the CCGT at 6 A.M. Run the coal units at 2000 MW during the off peak.

Option 2. Run one CT from the CCGT along with the steam turbine at minimum output during the off peak. Assume a start-up cost of \$10,000 for the remaining CT at 6 A.M. the following day. Run the coal units at an output of 1850 MW during the off peak.

The cost components of the two options are explained in the points below.

- The nuclear plant serves the same load requirement for both options. Accordingly, its cost does not affect the difference and is ignored.
- We consider cost differentials between 11 P.M. and 6 A.M. only, because the dispatch during all other hours is considered to be identical.
- The coal units' dispatch for Option 1 is identical to Case 2 in the previous example: Venus 1 dispatches at maximum capacity, 1200 MW, while Venus II dispatches at 800 MW.

Table 4.4 Cost for Options 1 and 2

	Cost in $ Option 1	Cost in $ Option 2
Venus I	321,832	321,322
Venus II	264,858	226,120
Orion (CCGT)	25,000	11,550*G* + 10,000
Total	611,690	557,952 + 11,550*G*

G, gas price in $/MMBTU

- Applying the solution method of the preceding example to Option 2 provides the following dispatch of the coal units: Venus I dispatches at 1200 MW, while Venus II dispatches at 650 MW.
- The costs of a coal unit are found by using the unit's associated cost curve.
- For Option 1, the cost of the CCGT is the start-up cost.
- For Option 2, the per MWH cost of dispatching the CCGT is found by multiplying the fuel price with the heat rate. The total cost is found by multiplying the per MWH cost by the total production between 11 P.M. and 6 A.M. and adding the start-up cost of the CT that does not operate throughout the off-peak period. Note that VOM is ignored.

The results of performing these cost calculations for Options 1 and 2 are presented in Table 4.4. The variable *G* is the gas price in $ per MMBTU.

Equating the total cost of the two options yields the following:

$$611{,}690 = 557{,}952 + 11{,}550G$$

Solving for *G* gives $G = 4.65$ dollars per MMBTU.

Whenever *G* is below 4.65 $/MMBTU, it is more economical to select Option 2 and dispatch the CCGT throughout the night. As of the writing of this text, the gas price is far above levels that would justify Option 2.

EXAMPLE 5. ***Combustion Turbine***

Combustion turbines are normally dispatched only during on-peak periods. Typically, because of high fuel costs, they are among the least economical alternatives for servicing load. When called upon to meet peak loads, CTs must be dispatched in accordance with their operational constraints.

4.5.5 Some Additional Features

Below are some additional interesting features of least-cost dispatch.

- During the morning shoulder hours there may be overproduction as units ramp up to meet on-peak loads. In the wintertime this occurs around 4 A.M. to 7 A.M. so that the early morning peak load can be served. In other months this can occur between 5 A.M. and 8 A.M.

- Often, a portion of hydro capacity is set aside for two ancillary services, regulation or spinning reserve. This is because hydro units are the most flexible and have the quickest response times to changes in grid conditions. Similarly, steam units often run at their optimal heat rate just below maximum capacity, allowing the difference to be used for spinning reserve. The next section addresses ancillary services.
- Wind-powered generation is dispatched as available. Planners must develop a production forecast in coordination with weather forecasters. The production forecast is incorporated into the output as a constraint. Wind power increases ancillary service requirements because of the uncertainty of the output. The next section incorporates wind power constraints.

4.6 LEAST-COST DISPATCH IN A SINGLE NODE WITH SPINNING RESERVE AND REGULATION

This section augments the constraint set in the previous section to ensure adequate spinning reserves in the dispatch. Recall that spinning reserve requirements are established in line with regulatory practices. In many cases the spinning reserve requirement is equal to the capacity of the largest unit dispatched, while regulation is a percentage of the forecasted load.

4.6.1 Assumptions

- Transmission charges and losses for all resources are negligible.
- There are no imports or exports between the control region and any of its neighbors.
- We ignore all network requirements except spinning reserve and regulation, e.g., standing reserve, voltage support are not considered.
- An hourly forecast is used to set the dispatch. The power consumption within each hour is constant.
- The fuel costs required to bring a unit up to minimum power are all contained in the start-up cost, while the fuel costs accrued during ramp-down from the minimum power threshold to zero are included in a shutdown cost.

4.6.2 In Words

- Objective

  The objective is to minimize the total cost of the dispatch. The total cost includes fuel costs, start-up costs, shutdown costs, and VOM charges. The decision variables available for attaining this objective are the output on each unit.

- Constraints
  1. The total power output must match the load forecast. The total output includes wind-generated resources. There is little control over the output from wind-generated resources; a forecast of output is used for planning purposes. A dumping term may be added to the load, indicating the amount of power that can be sent to ground.
  2. Each unit either provides no output or provides output bounded by minimum and maximum MW values.
  3. Each unit has a maximum ramp-up rate.
  4. Each unit has a maximum ramp-down rate.
  5. Once a unit is down, the unit has a minimum turnaround time.
  6. Once a unit is on line, the unit has a minimum on time.
  7. Units accrue a start-up cost for each start-up and a shutdown cost for each shutdown.
  8. The hydro facility has a maximum total energy availability.
  9. Regulation requirements must be met by units with the necessary operational capabilities.
  10. Spinning reserve requirements must be met.

As previously noted, systems with substantial wind resources require higher levels of regulation and spinning reserves than systems without wind resources.

4.6.3 In Equations

- Objective

$$\underset{\text{output}_{jh}}{\text{Min}} \sum_{h=1}^{48} \sum_{j=1}^{U} \text{cost}_j(\text{output}_{jh}) + u_{j,48}\text{SU}_j + \text{d}_{j,48}\text{SD}_j$$

  - $\text{cost}_j(\text{output}_{jh})$ is the cost incurred by unit j during hour h while producing output_{jh} MWH.
  - SU_j is the cost of a single start-up for unit j.
  - SD_j is the cost of a single shutdown for unit j.
  - $u_{j,48}$ is the total number of start-ups that occur over the time horizon.
  - $d_{j,48}$ is the total number of shutdowns that occur over the time horizon.
- Constraints
  1. $L_h \le \sum_{j=1}^{U} \text{output}_{jh} + \text{output}_{wh} \le L_h + \text{dumping}_h$ for each h
     - output_{wh} is the forecasted wind production for hour h.
     - dumping_h is the amount of energy that can be safely dumped in hour h.
  2. $\text{output}_{jh} = 0$ or $m_j \le \text{output}_{jh} \le M_j$ for each pair jh
  3. If $\text{output}_{jh} - \text{output}_{j,h-1} > 0$, then $\text{output}_{jh} - \text{output}_{j,h-1} < \text{ru}_j$
  4. If $\text{output}_{j,h-1} - \text{output}_{jh} > 0$, then $\text{output}_{j,h-1} - \text{output}_{jh} < \text{rd}_j$

5. If $\text{output}_{jh} > 0$ and $\text{output}_{j,h+1} = 0$, then $\text{output}_{j,h+k} = 0$ for all $k \leq \text{dt}_j$
6. If $\text{output}_{jh} = 0$ and $\text{output}_{j,h+1} > 0$, then $\text{output}_{j,h+k} > 0$ for all $k \leq \text{ut}_j$
7.

a. $u_{j,0} = 0$. If $\text{output}_{j,h-1} = 0$ and $\text{output}_{jh} > 0$, then $u_{jh} = u_{j,h-1} + 1$; otherwise, $u_{jh} = u_{j,h-1}$
b. $d_{j,0} = 0$. If $\text{output}_{jh} = 0$ and $\text{output}_{j,h-1} > 0$, then $d_{jh} = d_{j,h-1} + 1$; otherwise, $d_{jh} = d_{j,h-1}$

8. $\sum_{h=1}^{48} \text{output}_{jh} \leq w_j$ for all hydro facilities.

9.

a. $\sum_{k=1}^{R} M_k - \text{output}_{kh} > \text{RU}_h$

b. $\sum_{k=1}^{R} \text{output}_{kh} - m_k > \text{RD}_h$

- the index k ranges over all sets of units that are on line and are capable of providing regulation services.
- RU_h is the regulation up requirement for hour h.
- RD_h is the regulation down requirement for hour h.

10. $\sum_{j=1}^{U} M_j - \text{output}_{jh} > S_h$ for each h.

- S_h is the spinning reserve requirement for hour h.

4.7 LEAST-COST DISPATCH IN A NETWORK

We continue to add operational characteristics into the least-cost dispatch problem. In this section, network constraints are considered. There are two phases for determining least-cost dispatch within a constrained network. The first phase is to set a least-cost dispatch across a network. The dispatch solution to this set of equations may not be feasible; requirements to operate the transmission grid as described in Chapter 3 may not be met. Accordingly, operational feasibility of the solution must be verified. This is accomplished in a second phase by solving a problem known as the power flow problem.

There is an iterative process between the first and second phases whereby constraints in the least-cost dispatch problem are altered as determined by the solution to the power flow problem. If the solution to the power flow problem demonstrates feasibility, then the least-cost dispatch problem is finished. Otherwise, additional constraints must be incorporated into the original least-cost dispatch problem to force it toward feasibility. The problem is presented for a grid in which exports and imports may be considered across or within a control area.

4.7.1 Assumptions

- The network may be disaggregated into nodes with power flow constraints between the nodes. There is no unique way to split up a network into nodes; these do not follow strict definitions. Instead, they are a useful concept that assists in finding feasible solutions. Planners may define them geographically based on their knowledge of the transmission system. Figure 4.12 illustrates the division of a single control area into two nodes.

Within the diagram, power transfer capacity from one node to another is represented by a line with an associated power constraint. The line does not represent a transmission line. Indeed, there are typically many transmission lines between nodes. Instead the line represents the transfer capability of the transmission network based on power flow studies. In this diagram the transfer capability between nodes is symmetric; however, it may be that transfer capabilities may differ directionally. In such an instance two lines between nodes are required. The diagram also provides the customer's real demand along with available generation capacity within each node. This is not a detailed grid map as that provided in Chapter 3. Indeed, interconnection buses and distribution stations are all aggregated within the nodes, and reactive power requirements are not provided.

- There are n nodes. Each node has an associated generation stack and load requirement.

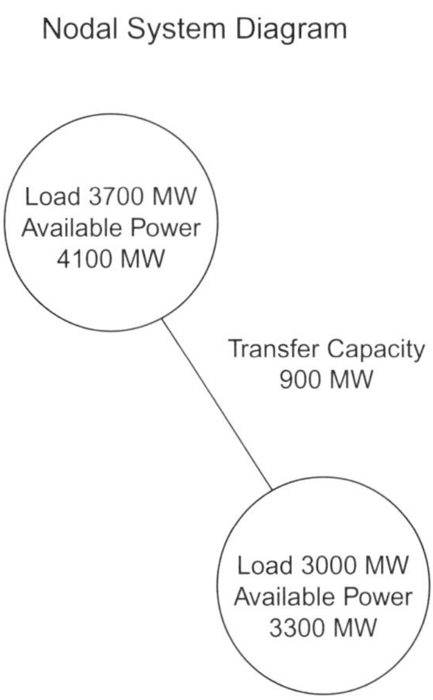

Figure 4.12

- There are no transmission fees.
- For each unit, transmission losses within the node where that unit resides are a percentage of the unit's output. This percentage is the same for all units. There are additional transmission losses between nodes.
- The power flow between nodes is constrained by the grid. Transmission flows between the nodes and neighboring control areas are prescheduled, and there are feasible solutions that allow for these power flows.
- An hourly load forecast is used to set the dispatch. The power consumption within each hour is constant.

4.7.2 Phase 1: Least-Cost Dispatch Problem in Words

- Objective: The objective is to minimize the total cost of the dispatch. The total cost includes fuel costs, start-up costs, shut-down costs, and VOM charges. The decision variables available for attaining this objective are the output on each unit.
- Constraints

  1. Supply and load must be balanced for each node and each hour. The total power supply within a node is the delivered power of units operating within that node plus the total delivered imported power from other nodes minus the total exported power from that node. Delivered power is power reduced by transmission losses. Imported and exported power within the control area is a decision variable.
  2. The imports and exports with neighboring control areas are fixed by contract. This occurs if the network includes control areas of different utilities and there are interutility contracts.
  3. There is a power transfer limit between nodes.
  4. Each unit either provides no output or provides output bounded by minimum and maximum MW values.
  5. Each unit has a maximum ramp-up rate.
  6. Each unit has a maximum ramp-down rate.
  7. Once a unit is down, the unit has a minimum turnaround time.
  8. Once a unit is on line, the unit has a minimum on time.
  9. Units accrue a start-up cost for each start-up and a shutdown cost for each shutdown.
  10. The hydro facility has a maximum total energy availability.
  11. Regulation requirements must be met by units with the necessary operational capabilities.
  12. Spinning reserve requirements must be met.

4.7.3 Phase 1: Least-Cost Dispatch Equations

- Objective

$$\min_{\text{output}_{jkh}} \sum_{h=1}^{48}\sum_{k=1}^{N}\sum_{j=1}^{U_k} \text{cost}_{jk}(\text{output}_{jkh}) + \text{u}_{jk,48}\text{SU}_{jk} + d_{jk,48}\text{SD}_{jk}$$

- Constraints

1. for all k,h

$$L_{\text{kh}} = \sum_{j=1}^{U_k}(1-\text{Tloss}_k)\text{output}_{kjh} + \sum_{i=1,\, i\neq \text{k}}^{N} \text{MWEx}_{ikh} - \sum_{i=1,\, i\neq \text{k}}^{N} (1-\text{Tloss}_{ki})\text{MWEx}_{kih}$$

- MWEx_{ikh} are additional decision variables representing exports from node i to node k.
- Tloss_k represents the transmission loss factor for node k. It is assumed that the loss factor is the same for all generation within a given node.
- Tloss_{ik} represents the transmission loss factor for transmitting power from node i to node k.

2. For each hour h and set of nodes ik:

$$\text{Contract}_{ikh} \leq \sum_{i=1,\, i\neq \text{k}}^{N} \text{MWEx}_{ikh}$$

- Contract_{ikh} represents any contractual agreement to deliver power from node i to node k for hour h.

3. for each $i \leq n$ and $k \leq n$, $\text{MWEx}_{ik} \leq \text{Tlimit}_{ik}$

- Tlimit_{jk} represents the limitation of power flows due to transmission constraints.

4. $\text{output}_{jkh} = 0$ or $m_{jk} \leq \text{output}_{jkh} \leq M_{jk}$ for each jkh

5. If $\text{output}_{jkh} - \text{output}_{jk,h-1} > 0$, then $\text{output}_{jkh} - \text{output}_{jk,h-1} < u_{jk}$

6. If $\text{output}_{jk,h-1} - \text{output}_{jkh} > 0$, then $\text{output}_{jk,h-1} - \text{output}_{jkh} < d_{jk}$

7. If $\text{output}_{jkh} = 0$ and $\text{output}_{jk,h+1} = 0$, then $\text{output}_{jk,h+H} = 0$ for all $H \leq dt_{jk}$

- dt_{jk} is the minimum down time for unit j associated with node k.

8. If $\text{output}_{jkh} = 0$ and $\text{output}_{jk,h+1} > 0$, then $\text{output}_{jk,h+H} > 0$ for all $k \leq ut_{jk}$

9. $u_{jk0} = 0$. If $\text{output}_{jk,h-1} = 0$ and $\text{output}_{jkh} > 0$, then $u_{jkh} = u_{jk,h-1} + 1$; otherwise, $u_{jkh} = u_{jk,h-1}$
$d_{j0} = 0$. If $\text{output}_{jkh} > 0$ and $\text{output}_{jk,h-1} = 0$, then $d_{jkh} = d_{jk,h-1} + 1$; otherwise $d_{jkh} = \text{d}_{jk,h-1}$

10. $\sum_{h=1}^{48} \text{output}_{jkh} \leq w_{jk}$

- output_{jkh} represents the output of hydro facility j in region k during hour h.

11.

a. $\sum_{j=1}^{R(k)} M_{jk} - \text{output}_{jkh} > \text{RU}_{kh}$

b. $\sum_{j=1}^{R(k)} \text{output}_{jkh} - m_{jk} > \text{RD}_{kh}$

c. $\sum_{k=1}^{n} \sum_{j=1}^{R(k)} M_{jk} - \text{output}_{jkh} > \text{RU}_h$

d. $\sum_{k=1}^{n} \sum_{j=1}^{R(k)} \text{output}_{jkh} - m_{jk} > \text{RU}_h$

- Constraints a and b are local requirements within node k and apply to each node. The index ranges over all sets of units that are on line and are capable of providing regulation services to node k. This index set is different from the one above that includes all units. Accordingly, the upper bound is designated by $R(k)$ as opposed to $S(k)$.
- RU_{kh} is the regulation up requirement for hour h in node k.
- RD_{kh} is the regulation down requirement for hour h in node k.
- Constraints c and d are global requirements across all nodes. The index j ranges over all units capable of providing regulation services.
- RU_h is the global regulation up requirement for hour h.
- RD_h is the global regulation down requirement for hour h.

12. Spinning reserverve

a. $\sum_{j=1}^{SR(k)} M_{jk} - \text{output}_{jkh} > \text{SR}_{kh}$

b. $\sum_{j=1}^{n} \sum_{k=1}^{\text{SR}(k)} \text{output}_{jkh} - \text{m}_{jk} > \text{SR}_h$

- Constraint a provides local requirements within node r and applies to each node. The index j ranges over all sets of units that are on line and are capable of providing regulation services to node r. This set is different from the previous set of units.
- SR_{kh} is the spinning reserve requirement for hour h in node k.
- Constraint b provides global requirements across all nodes. The index j ranges over all units capable of providing spinning reserve services.
- SR_h is the global regulation up requirement for hour h.

Remarks

- The decision variables enter into both the objective function and the first constraint. MWEx_{ikh} is a decision variable for the quantity of power to be exported from node i to node k.
- This approach does not model the transmission network. Transmission losses in this set of equations are approximations and may be altered by the output of the power phase problem of Phase 2.

- The last equation, Equation 11, designates ancillary service (AS) and regulation requirements. It may be possible for the nodes to share these requirements. In this case, n represents the number of regions that require independent AS and regulation requirements and is smaller than N.

4.7.4 Phase 2: Checking for Feasibility and Ensuring Reliability: Power Flow Analysis

The least-cost dispatch established in Phase 1 does not necessarily provide a feasible solution. The nodal analysis is not complete enough to determine whether or not the grid can support power flows as established by the dispatch. A complete analysis must determine several factors, including:

1. Current quantities through transmission lines to ensure that they are within limits
2. Voltage magnitudes at load buses to ensure that voltage is sufficient
3. Reactive power output from generators to ensure that total power output is within generator specifications
4. Phase angle differentials between buses to ensure that the differences are acceptable (recall that the grid is unstable when phase angle differences become too large)
5. Robustness of the system to ensure that the grid can withstand contingencies

To ensure that the solution is feasible the solution must be analyzed by a power flow analysis. In a complete power flow analysis, all components of the grid are mapped, generators, buses, transmission lines, and control devices. Real and reactive load requirements at each load bus are input from a forecast. Supplies at the supply buses are input from the least-cost dispatch problem of Phase 1. Equations that model power flows are then solved for these inputs. Finally, contingencies including unit outages and transmission line outages are simulated. With this rigorous approach, the concerns above are addressed.

If the least-cost solution is not feasible because of a problem with any of the areas identified above, the dispatch of Phase 1 must be altered to address the concern. For example, a unit that is not a solution to the least-cost dispatch problem may be dispatched to provide reactive power. This adjustment is input in the form of a constraint that is added to the nodal problem, and the nodal problem is once again solved. An iterative process between the nodal problem and flow analysis results in a least-cost feasible solution.

4.7.5 The Power Flow Problem and System Feasibility

Determination of power flow through the grid given demand and output variables is known as the power flow problem. The problem of interest is to arrive at a description of the operating conditions of the grid under the dispatch that results from Phase 1. This is expressed more concretely as follows.

The following variables are known:

1. Real and reactive load requirements at each bus connected to distribution lines
2. Real power output of all generators except one called the reference generator
3. The magnitude of the voltage at each generation bus (any bus where a generator ties into the grid)
4. The phase angle of a reference bus, the bus where the reference generator interconnects
5. The resistance, inductance, capacitance, and current limitations of all transmission lines
6. Capabilities of any supplementary equipment such as capacitors

The problem is to solve for the remaining variables that are required to describe the system.

1. The magnitude of the voltage at every load bus
2. The phase angle of the voltage at every bus
3. The reactive power output of each generator
4. The real power output of the reference generator
5. The current and power flows through the transmission lines

With a solution to the remaining variables a complete description of current and power flows through the grid is available. We do not present the mathematical formulation of the power flow problem as it is beyond the scope of this text. A general treatment of the problem is found in standard power systems analysis texts such as that of Bergen.

Remarks on the Power Flow Analysis

- Conditions for the power flow problem are set to match predicted power demands at a single instant of time. The solution provides a snapshot of events for the same instant in time. One can change these conditions to determine whether the outcome of Phase 1 is sufficiently robust to address contingencies. As examples, one could simulate a line outage by removing a line and resolving the problem, increase the reactive power requirements, or simulate a unit outage and bring on the units that are identified for primary and secondary reserves.
- The undetermined real power level at the reference bus provides an additional degree of freedom that is required to solve the power flow equations. The need comes from the fact that system losses are not known a priori; the extra degree of freedom is required so that the solution balances supply with demand and

losses. This mathematical requirement of an additional degree of freedom parallels actual grid conditions; as explained in Chapter 3, there are power plants that are designated to adjust their output to load requirements.

Final Remarks

- This section illustrates the link between grid management and dispatch of generation. There is no natural decoupling of the problems; instead, grid management and economic dispatch is an integrated problem.
- The approach requires a process for adjusting or augmenting constraints in Phase 1 based on the results of Phase 2. This is still very much a research topic.
- Many experts disagree with the use of a network in Phase 1. Their preferred approach is to use the problem as posed in the previous section for Phase 1 and place constraints on individual units based on results of Phase 2.

4.8 REAL TIME

Real-time operations are discussed in Chapter 3. This section briefly summarizes the discussion of Chapter 3 and then places real-time operations within the context of economic dispatch. The role of the real-time dispatching team is to monitor system parameters and modify day-ahead planning to accommodate actual conditions. The real-time dispatch team is part of the grid operations. The real-time dispatching team must be cognizant of the same issues that day-ahead planners address: cost and reliability.

As noted in Chapter 3, information systems provide dispatchers with real-time system outputs, as well as network imports and exports. Real-time dispatchers monitor the following variables.

1. Output of each unit
2. Unit status
3. System imports and exports
4. AS provisions from each unith
5. System frequency
6. System status, transmission lines, capacitors

Additionally, the dispatching team includes a forecaster who provides updates to the remainder of day load forecast. This information allows real-time dispatchers to run systems that resolve the economic dispatch problem and reset dispatch schedules throughout the day.

Under normal conditions operations do not deviate considerably from the day-ahead least-cost dispatch. There are some exceptions:

- Load is considerably different from the day-ahead forecast.
- Failure of a grid component such as transmission line, transformer, voltage control device

- System imports and exports deviate from schedule because of problems in neighboring control area
- Unforeseen power flows due to interconnect operations
- Failure of a unit

Real-time dispatchers must be prepared to respond appropriately and economically to all of these contingencies, as described in Chapter 3.

Chapter 5

Long-Term Utility Planning

This chapter formulates the fundamentals of long-term planning and operations within a utility environment. Long-term planning addresses the economic selection of generation and transmission additions necessary to meet projected load requirements.

Section 5.1 begins with a description of project development. Planning activities support the development process, and an understanding of the development process assists in providing relevant information. In Section 5.2 we provide an overview of the planning process. Next, an approach to long-term demand forecasting is given in Section 5.3. Finally, we address supply-side planning and formulate the problem of making optimal choices in Sections 5.4 through 5.7. As with Chapter 4, we layer the complexity of supply-side planning section by section. We begin with a simple approach to generation capacity planning within a single control area and gradually build toward an integrated approach that couples transmission additions with generation additions. Section 5.8 ends the chapter with a brief description of determining reserve requirements.

The planning studies presented here are generic studies with the objective of providing guidance for portfolio development. The portfolio referred to is the set of generation and transmission resources that is used to service load. The output of the studies informs decision makers of the quantities of technologies that are necessary to meet future demand. They answer such questions as, "How many megawatts of combustion turbines versus combined cycles are required to match future load most economically?". The studies do not address the merits of a specific project. With such studies, project developers can then begin to look for sites and make preparations to construct new generation.

5.1 PROJECT DEVELOPMENT

This section provides a brief introduction to project development. The purpose is to understand the context of the studies that we present later in the chapter. Since the

Electric Power Planning for Regulated and Deregulated Markets. By Arthur Mazer

studies are meant to provide guidance to project developers, one must understand the issues that developers confront. Then it is possible to provide them with relevant and timely information.

5.1.1 Generation Project Development

Project development encompasses the process of constructing a new generating facility or transmission equipment from the conceptual phase until the moment when the facility is operating. Planners must have recommendations ready for project developers to complete construction in time to meet customer load requirements.

Aside from meeting the requirements of a baseline forecast, timelines are also critical for contingency planning. Actual demands may vary from baseline forecasts, and planning must address this contingency. The timeline of construction provides an indication of how quickly one can respond to unforeseen load demand.

Generation project development encompasses six related activities: siting, permitting, grid assessment, fuel delivery assessment, financing, and construction.

5.1.2 Siting

Often utilities take ownership of sites for new construction years before construction is anticipated. There are several issues that a utility must consider in selecting sites. Among these are the site's proximity to an interconnection bus, the impact of a project on the transmission system, access to fuel transportation facilities, property expense, and permitting requirements.

5.1.3 Permitting

Permitting is the process of receiving the authority to construct and operate from the various regulatory agencies that authorize projects. There are many different agencies overlooking different aspects of the project. At the state level there is a utility commission that authorizes each project. A utility must demonstrate to the commission the need, cost-effectiveness, and viability of the project. There are also local zoning agencies that must approve projects. Finally, there are state and federal agencies that oversee environmental issues. Each agency has its own application and processing requirements that a utility must satisfy in order to obtain permits. An important distinction is the right to construct versus the right to operate a unit. It is possible that a project meets the criteria necessary to allow construction. However, there might be air quality constraints that prohibit operations of the unit. In such a case the agencies authorizing construction would give their approval; however, agencies overseeing emissions control would withhold their permits. Needless to say, a well-functioning utility commission would not approve of such a project.

5.1.4 Grid Assessment

A system impact analysis that determines the impact of a project on the grid must be performed before construction. The system impact study determines whether or not the project affects grid operations and equipment. Many projects have an adverse impact; for example, power flows across a transmission line may increase beyond the line's tolerance. The study includes remedial actions to ensure reliable grid service. Remedial actions include the addition of grid equipment such as transmission lines, capacitors, circuit breakers, and transformers. The cost of system upgrades is often quite substantial. These costs are included in the filing that the utility submits to the utility commission for approval. Grid additions also require approval from local, state, and federal agencies. They are part of the package that must be put forward for permitting approval.

5.1.5 Fuel Delivery

An additional concern is ensuring an adequate transport system for fuel delivery. In the case of a gas-powered project, this requires a study to determine the project's impact on local gas pipeline flows. Upgrades to the pipeline may be necessary to ensure gas delivery. Additional pipeline construction is also required from the gas pipeline system to the burner tip of the proposed project. As with transmission upgrades, pipeline upgrades are considered part of the project and must be vetted through the project approval process.

5.1.6 Financing

Utilities obtain financing of a project before construction. Banks view such projects very favorably once they have passed the permitting stage. As part of a utility's application with the utility commission there is a mechanism for recovering the cost of the project through regulated rates to the end customer: rate-based cost recovery. Banks find this provision particularly attractive, and utilities can generally get favorable lending terms.

5.1.7 Construction

Construction requires the management of supplies, materials, labor, and expertise necessary to complete the project and bring it on line. The process proceeds from ordering supplies and materials through construction initial testing and final testing. Typically, a utility does not use its own personnel for actual construction work; instead, utilities contract a specialized construction company along with many subcontractors.

The timeline for projects differs with technology as well as local permitting regulations and processes; the permitting process in Alabama is far less onerous and accordingly much swifter than the permitting process in California. Table 5.1 presents an overview of the development process for a combustion turbine (CT) within California.

Table 5.1 Timeline, Permitting, and Construction of CT

Timeline		
Activity	Start time beginning of month	Completion time end of month
Siting	0	6
Gas pipeline impact study	5	6
System impact study grid	5	6
Permitting process	6	18
Financing	12	18
Construction	18	28
Testing	28	29
On line	29	On line 20 to 30 years
Permitting process		
Agency	Responsibility	
California Energy Commission	State land use	
Air Pollution Control District	Emissions permitting	
Regional Water Control Board	Waste water discharge	
California Coastal Commission	Coastal development planning	
US Environmental Protection Agency	Emissions permitting	
California Department of Fish and Game	Reviews impact on fish and wildlife, endangered species Consultation	
State Land Commission	Approves use of state lands	
California Public Utilities Commission	Approves cost recovery through regulated rates	
Construction activities		
Preconstruction		
Design		
Contracting for Labor		
Contracting for Material		
Construction		
Site preparation (clearing,		
water connections		
gas pipeline interconnection		
Installation by component (air intake, combustion chamber, turbine, condenser, generator)		
Interconnection to grid (transformer, radial line)		
Grid upgrades (circuit breakers, transformer upgrades, transmission line upgrades)		

5.1.8 Transmission Project Development

There are two approaches to enhancing the transmission system. The first is to perform an upgrade as a stand-alone project. The second is to perform the upgrade in association with the addition of a new generation unit. The first approach, a transmission upgrade as a stand-alone project, proceeds in a manner similar to the development of generation projects. Four of the six activities must be addressed: siting, permitting, financing, and construction.

A significant cost for constructing new transmission lines is the acquisition of land along the proposed transmission line's pathway. Siting of power lines must ensure that the benefit of additional power deliveries via the proposed line offsets the land acquisition costs as well as the cost of construction.

The permitting process for transmission lines is arduous. As with power plants, there are various local, state, and federal agencies that must provide permits before the project can proceed.

The financing of a transmission project is similar to that of a power plant. Once a utility commission authorizes the project, costs can be recovered through the regulated customer rate mechanism. As with generation, banks view such projects favorably and financing is forthcoming.

Transmission projects are also initiated as remedial actions that are noted in a system impact study for a power plant proposal. As noted above, a new power plant influences power flows along the grid. A power flow analysis as described in Chapter 4 is performed to determine alterations in power flows. From this study one can determine additional transmission carrying capacity required to support the additional power flows. Typically, the most economical alternative for upgrading the transmission carrying capacity is to resize the existing lines of the transmission system; however, in some cases the building of new lines may be justified. Either way, this is an expensive undertaking. It is common that upgrades are substantially oversized to incorporate needs beyond the addition of the associated power plant.

The diagram in Figure 5.1 illustrates the following situation. Suppose that G10 is a proposed plant with interconnection bus B3. Suppose also that power flow studies indicate that the addition of G10 would cause an additional 100 MW of power to flow along the transmission line from bus B3 to bus B4. Suppose further that this line is rated to carry only an additional 70 MW beyond its current use. Remedial action would require an addition 30 MW of line capacity before starting up the proposed power plant. Given the expense of upgrading the line, it would be foolish to increase the line capacity by only 30 MW. Instead, in such situations it is common to increase line capacity by several hundreds of MW in anticipation of future needs. Utility commissions recognize the economic advantage of this approach and approve of such projects.

The example once again illustrates the integrality of generation and transmission systems. There are economic advantages in a unified planning approach.

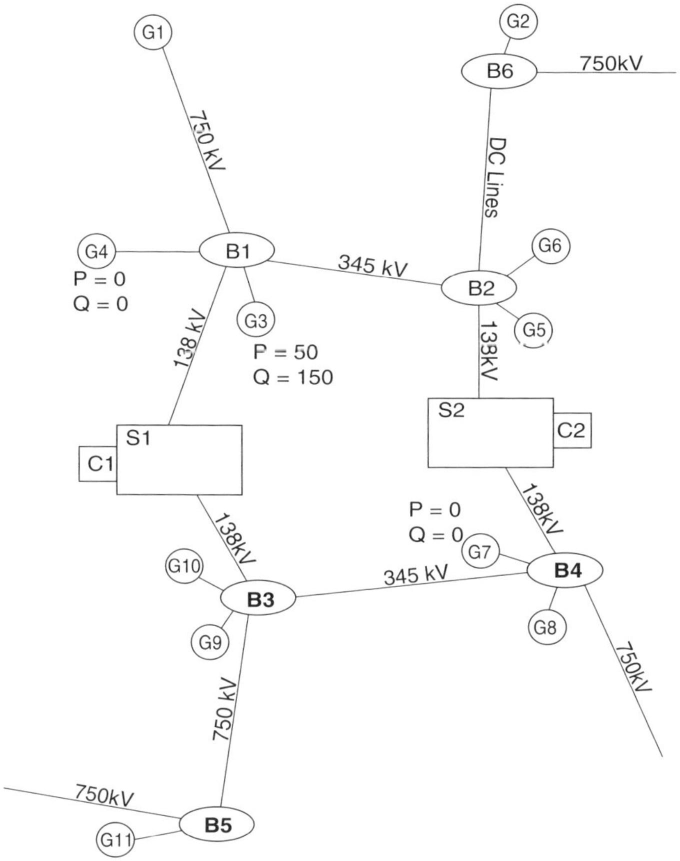

Figure 5.1

5.2 THE PLANNING PROCESS

We restrict our discussion of the planning process to providing guidance for selecting generation and capacity additions as opposed to assessing the merits of a particular project. This section gives an overview of the planning process in a list format. Details of the process leading to quantitative analysis are provided in subsequent sections.

Step 1. Pose an objective.

At this stage, it is not necessary to mathematically formalize an objective function along with constraints. However, it is necessary to provide an objective that frames other considerations. Details of the optimization problem are determined at a later stage. In words, the objective is to minimize the cost to service load over a relevant time horizon. The costs include all dispatch costs, as presented in Chapter 4, all capital expenses for new equipment, and all fixed costs required to keep a facility available for use.

The time horizon of the study is necessarily much longer than the development time for new projects. This is because one wishes to investigate the cost-effectiveness of new projects over a significant portion of their lifetime. Since the lifetime of a new plant is 30 years and development times may be 6 years, it is not out of line to initiate a 36-year study.

Step 2. Develop a complete list of factors that affect the cost of servicing customer load.

Below is a list.

1. Gas prices (forecast)
2. Coal prices (forecast)
3. Other fuel prices (forecast)
4. Customer load (forecast)
5. Primary and secondary reserve requirements (forecast)
6. Operational characteristics of current power units (forecast)
7. Current fuel contracts (actual)
8. Cost of construction for different generation technologies (forecast)
9. Cost of construction for different locations (forecast)
10. Cost of interconnecting to the grid (forecast)
11. Time of construction (forecast)
12. Operational characteristics of new construction (forecast)
13. Future maintenance cost of current units (forecast)
14. Maintenance cost of new generation construction (forecast)
15. Costs to mitigate environmental regulation (forecast)
16. Fixed costs required to keep a unit in operational order, e.g., personnel, building maintenance, others (forecast)
17. Future configuration of the grid (forecast)
18. Cost of construction for new transmission (forecast)
19. Maintenance cost of grid (forecast)
20. Environmental costs (forecast)
21. Discount rate (determined by interest rate and return)

Note that, excluding points 7 and 21, all factors are forecasts. Factor 21 is the discount rate. The present value of a future cash flow diminishes as time of the cash flow becomes more distant; the discount rate provides a way to discount future cash flows. A further discussion of the discount rate is given in Chapter 9.

The list is quite extensive. In this chapter we provide details of load forecasting and assume that forecasts for the other factors are available.

Step 3. Mathematically formalize the optimization problem and establish an algorithm that optimizes the objective function while considering the factors in Step 2.

The statement of the optimization problem must be formalized. This includes a concrete description of the objective function along with a presentation of the constraints.

Additionally, a solution method for selecting the optimal choice variables must be determined. Both the formalization of the problem and the solution method are within this single step because each element of this pair affects the other. There may be several iterations between the problem statement and the solution method before arriving at a satisfactory pair.

This step must address the following factors.

- Details of the choice variables such as size, location, and type of new generation and transmission lines as well as retirements
- The method of calculating costs of different choice variables
- A method for arriving at the optimal solution

Subsequent sections of this chapter address different approaches for this step.

Step 4. Prepare forecasts for all the factors in Step 2 and transform them into inputs required in Step 3.

This is a very difficult task and is ultimately what drives the result. The list of forecasting requirements is very extensive. Additionally, one must transform the raw data of the factors present in Step 2 into input parameters and cost values that are consistent with the solution method selected in Step 3.

Step 5. Execute the algorithm and use the outcome to prepare recommendations for new construction.

5.3 LONG-TERM LOAD FORECASTING

The previous section provides a long list of forecasts that affect the generation resource selection. It is not within the scope of this text to present methodologies for forecasting each element of the list. However, in this section we provide a brief glimpse into long-term load forecasting.

Before proceeding we note that the requirements for load forecasts are stringent compared to those of other industries. It is insufficient to forecast energy consumption by year, or even by day. More resolution is required in order to determine the type and quantity of technology that most economically matches load; peaking load, intermediate load, and baseload load are optimally matched to peaking, intermediate, and baseload supply. The exact resolution of the load forecast depends on the method that is used for selecting capacity. In this section, we consider hourly load forecasts.

Long-term forecasting addresses the annual changes in demand that reflect changes in population and GDP. Although weather is a major driver for short-term forecasting with an explicit role, it plays a background role in long-term forecasting. For a base case scenario weather is not assumed to change dramatically from year to year; base case weather assumptions apply. Two methods for addressing annual changes in demand follow.

Method 1. Simple Method

Determine the historical annual rate of growth in total energy consumption and apply this growth rate to recent data in order to project future consumption. The equation is as follows:

$$L(h_{jk}) = L(h_{j0}) \times (1 + r)^k$$

where

- $L(h_{jk})$ represents the load for hour j in year k.
- r is the annual rate of growth in the load.
- Year 0 represents the previous 12 months.

There are some shortcomings that must be addressed. First, the entire projection is based upon year 0, and there may be anomalies in year 0 due to weather or economic factors. Second, the model assumes that the load shape is static. This is typically not the case; in many areas load profiles have become more uneven. The demand difference between the hour of greatest demand and the hour of least demand has grown; this trend is not reflected in the formula.

Method 2. Regression Against GDP and Population

Run a regression of total annual load against population and GDP. Input a forecast of population and GDP into the regression formula to forecast total annual load demand. Profile hourly shapes, using the previous year's data. The equations for this method follow.

$$L_k = a(\text{GDP}_k) + b(\text{Pop}_k) + c$$

where

- L_k is the total demand for year k,
- GDP_k is the forecasted gross domestic product for year k.
- Pop_k is the forecasted population of year k.
- a, b, and c are regression coefficients from historical data.

Once the total demand is forecasted, the demand is shaped into an hourly price as follows:

$$L_{jk} = S_{jk}L_k$$

where

- L_{jk} is the load for hour j of year k,
- S_{jk} is the shaping coefficient for hour j of year k.

Forecasting the shaping coefficient is very difficult. Although it is possible to identify factors that affect hourly profiles, it is difficult to forecast them. One approach to the problem is to segregate the forecast into different load types. Industrial load is forecast separately from residential and commercial load. While the shape of industrial load remains fairly constant over the years, the shape of residential load changes. Increased use of central air conditioning has caused greater unevenness in residential load shapes, with the difference between daytime and nighttime usage increasing. There are some forecasters that adopt the load shape by measuring the historical trend in the load shape's change and applying that trend throughout the forecast horizon. Other forecasters ignore a changing load shape and maintain a fixed shape throughout the forecasting time horizon.

Remarks

- There are several modifications of Method 1. For example, rather than applying the growth factor to year 0, one could model a base year forecast by applying the growth factor to past years' loads and averaging across the results.
- The forecast of Method 2 depends on the quality of the GDP forecast and the population forecast.

5.4 A SIMPLIFIED LOOK AT GENERATION CAPACITY ADDITIONS

This section formally introduces an optimization problem for capacity selection and provides a solution method; this is Step 3 of the five-step process introduced in Section 5.2. The approach makes many simplifications; by simplifying, it is possible to highlight critical issues that govern the decision making process. The section begins with the simplifying assumptions. The methodology is then explained and illustrated with an example. Afterwards a step-by-step approach for a solution is described, and then another example is presented. Finally, the problem is stated in a slightly more realistic way as a preparation for the next section. It should be noted that variable costs and unit dispatch follow the merit order dispatch described in Section 4.3, where units are stacked against load in merit order. Accordingly, this section is the long-term counterpart to Section 4.3.

5.4.1 Assumptions

Below is a list of assumptions that simplify the problem.

- There is annualized year-on-year load growth requiring annual additions of capacity.
- Capacity retirements are not considered.
- All transmission issues are neglected, including capacity additions, upgrades, and constraints.
- Costs of new construction depend on technology and are easily disaggregated into fixed and variable components.
- Units dispatch in merit order as described in Section 4.3. The assumptions of that section apply. In particular, temporal constraints such as ramp-up rate are not considered.

- The cost of fuel varies little throughout the year. Accordingly, approximating the variable cost of production by an annual per MWH number is reasonable.

5.4.2 Statement of the Problem

In this section the approach is to determine the incremental quantity of capacity by technology type that is necessary to meet the incremental load and reserve requirements of a specified year. The assumption is that the load requirements for the previous year are appropriately addressed by a well-designed generation portfolio; accordingly, we address incremental requirements. This is a tremendous simplification that allows us to arrive at an insightful solution technique. Repeating the approach for subsequent years allows one to develop a report of capacity addition requirements by year over the time duration of the study.

A formalization of the problem is presented along with the solution method. The formalization includes the objective function as well as the constraints. The presentation follows that of previous chapters; the problem is first described in words and then in equations.

5.4.3 In Words

- Objective

For a given year, minimize the total cost of capacity additions in MW by technology.

- Constraints
  1. Capacity constraints: There must be enough capacity additions to meet incremental load and reserve requirements.
  2. Development time constraints: Capacity may not come on line before the time required by the development cycle.

5.4.4 In Equations

- Objective

$$\underset{\mathrm{MW}_j^y}{\mathrm{Min}} \sum_{j=1}^{P} \mathrm{MW}_j^y(\mathrm{Fixed}_j^y + \mathrm{Variable}_j^y)$$

  - The superscript y represents the year of interest and begins with $y = 0$ as the current year. This is only an identifier and does not indicate the mathematical operation exponentiation.
  - MW_j^y represents the capacity additions in MW of technology j for year y.

  - Fixed_j^y and Variable_j^y represent the fixed and variable costs of technology j during year y.
  - There are P technology types.
- Constraints
  1. $\sum_{j=1}^{P} \text{MW}_j^y \geq \text{MW}^y$
     - MW_j^y represents the capacity associated with the decision variables.
     - MW^y represents the incremental capacity requirement for year y and is determined by the incremental load and reserve requirements.
  2. If $y < \text{DT}_j$, then $\text{MW}_j^y = 0$.
     - DT_j represents the development time of technology j.

Note that the optimization problem is defined over a single year. To perform a multiyear study, the problem is redefined and solved for each year of the study's time horizon. Also note that an alternative way to apply Constraint 2 is to eliminate any technologies that cannot be complete by year y from the selection set and only consider those that can be on line. In what follows, we adopt this approach.

5.4.5 Solution Method Explained and Illustrated

We next present the solution method for this problem. The idea is to make simplifying assumptions about the cost of each technology that allow one to plot the annual per MW cost as a function of time of operation. Cost curves for different technologies generally intersect as a function of the time in operation. Intersections denote a point of change in the selection decision. Matching the cost curves against the incremental load and reserve requirements for the specified year provides the optimal solution. We illustrate this technique with an example.

Consider two decision variables associated with two technologies, each having the following corresponding cost curves.

$$\text{Cost}_1(X) = 82{,}736 + 56.91X$$

$$\text{Cost}_2(X) = 63{,}270 + 79.50X$$

- X is the hours of operation.
- The units for cost curves are \$ per MW.

Figure 5.2 provides a plot of the cost curves.

Note that the cost curves intersect at the point $X = 862$ hours. Below 862 hours of operating time, Technology 2 is the optimal choice. Above 862 hours of operating time, Technology 1 is the optimal choice.

All that remains to solve the problem is to determine the annual incremental MW requirement for units that operate less than 862 hours and units that operate greater than 862 hours. Suppose that an incremental hourly load forecast is available. Each hourly number represents the incremental hourly energy requirement or the

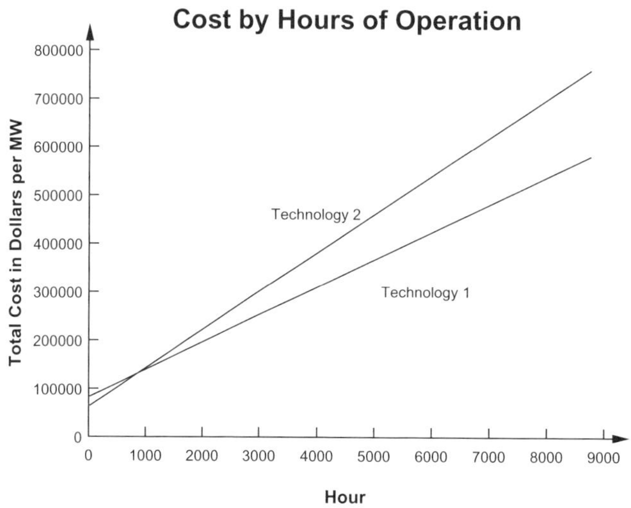

Figure 5.2

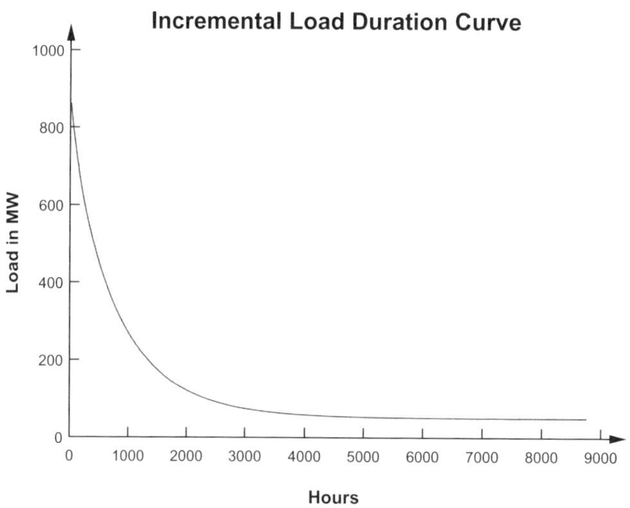

Figure 5.3

incremental hourly power requirement that is assumed to be constant over the given hour. We demonstrate the forecast on a graph known as the load duration curve, as illustrated in Figure 5.3. The horizontal axis of the graph represents hours of incremental load, and the vertical axis represents load requirement in MW. The graph is made by ranking the 8760 incremental hourly loads from high to low and then plotting them. (There are 8760 hours in a year.)

Note that at $X = 0$ the incremental load requirement is 908 MW, while at $X = 862$ hours the incremental load requirement is 208 MW. This means that 600 MW are

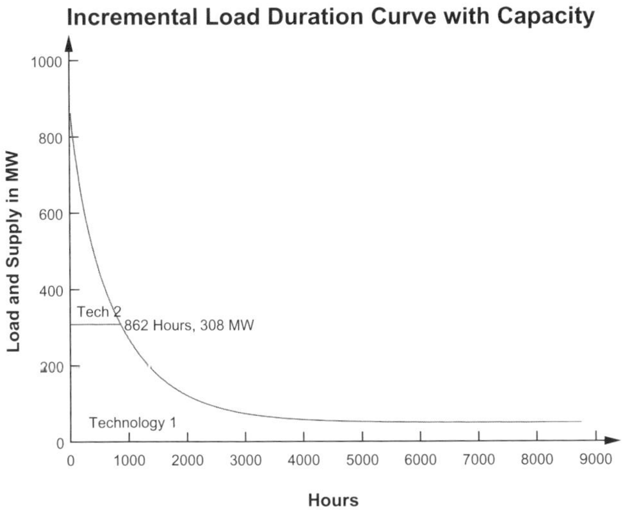

Figure 5.4

required to service the band of load that runs for less than 862 hours. The optimal technology choice for this band is Technology 2. The remaining 300 MW of requirement to service load is filled by Technology 1. The graph in Figure 5.4 illustrates the energy bands that are filled by the different technologies. If there is an additional reserve requirement, there should be a corresponding capacity addition from Technology 2.

While providing an outline of the solution, the explanation does not address details such as determining the incremental load and costs. These details are presented below.

5.4.6 Step-by-Step Method for Solving the Optimization Problem

Below is a step-by-step method for carrying out the solution technique that is illustrated in the preceding paragraph. Note that these steps are applicable to each year in the time horizon of the study. For example, if the wish is to develop a 7-year study that guides capacity additions, these steps would be carried out for each year of the study, providing results of desired generation additions by technology over 7 years.

Step 1. Prepare load forecasts and a load duration curve.

This step may be accomplished by following the process outlined in Section 5.3, ranking the loads from high to low and then plotting them in accordance with their rank.

Step 2. Prepare variable costs in $ per MWH for all current and planned generation.

Current generation is generation already on line. Planned generation may be generation under development that is due to come on line before the year under investigation or the result of the optimization problem applied to a preceding year. For example, if the result of the optimization problem for year 4 is that 400 MW of CCGTs should enter service, then the problem defined for years 5 and beyond incorporates those 400 MW of CCGTs as planned capacity. The units of the variable cost are $ per MWH. Multiplying by the capacity and number of hours in operation provides the energy cost for the year.

Although the energy cost fluctuates with fuel price and fuel prices are seasonal, for simplicity we consider fluctuations to be small. Accordingly, the variable costs are set on an annual basis.

Step 3. Prepare fixed and variable costs for all new technology types that comprise the selection variables in the objective function.

The fixed and variable costs allow for the development of cost curves with their associated break points. The fixed cost represents the cost of construction, while the variable cost represents the cost of operations.

Step 4. Determine the dispatch levels of the break points where the optimal choice among the technology selections transitions from one technology to another.

With the fixed and variable costs, the cost curves are plotted as a function of output. The points of intersection for these curves are the break points.

Step 5. Develop a merit order stack for all generation; this includes current generation, planned generation, and technologies that are under consideration for selection.

As described in Section 4.3, the merit order stack is obtained by ordering the generation units from lowest to highest by their variable cost; that is, the unit with the lowest variable cost receives the first position in merit order and so on.

Step 6. Stack the capacity of the current and planned generation units as well as the selection technologies into the load duration curve in merit order. The MW volumes for the selection choices are determined by the minimum of the volume that places the next technology selection at its break point or the volume that fills the total annual MW requirement including reserve.

The capacity is stacked across the annual load in the same manner as capacity is stacked across the daily load in Section 4.3.

Step 7. By defining the incremental load as the load that the selection technologies fill on being stacked under the load duration curve, the MW volumes selected in Step 6 provide the solution to the optimization problem.

The definition of incremental load steers the problem to the simple solution method. This simple approach demonstrates fundamental issues that arise in a more realistic setting. These issues are illustrated in the example below. After the example, a slight variant of this problem is presented; the variant does not impose the above artificial definition of an incremental load and leads to an improved solution.

EXAMPLE *Capacity Selection*

We apply the step-by-step solution method to an example. The study horizon for the example is year 5, and we indicate the impact of year 5's selection on year 6.

Step 1. Prepare load forecasts and a load duration curve.

Load forecasting is discussed in Section 5.3. Figure 5.5 provides the load duration curve along with reservation requirements for year 5 that we assume for this example.

Step 2. Prepare variable costs in $ per MWH for all current and planned generation.

Table 5.2 provides an assessment of installed capacity for year 5.

Step 3. Prepare fixed and variable costs for all new technology types that comprise the selection variables in the objective function.

We introduce three technology types: baseload, intermediate, and peaker. The baseload technology is a coal-fired steam generation plant with extensive emissions reduction technology. The full development timeline for this technology is 5 years, so this technology is not considered before year 6. The intermediate technology is a CCGT, and the peaker technology is a CT. The time of construction for both of these technologies is less than 4 years; accordingly, they are considered for both years 5 and 6. The variable components of these technologies are presented in Table 5.3.

The annual per MW fixed cost consists of the total annual payment to finance construction. For a given technology, the fixed cost of construction is taken from Section 2.3.8.

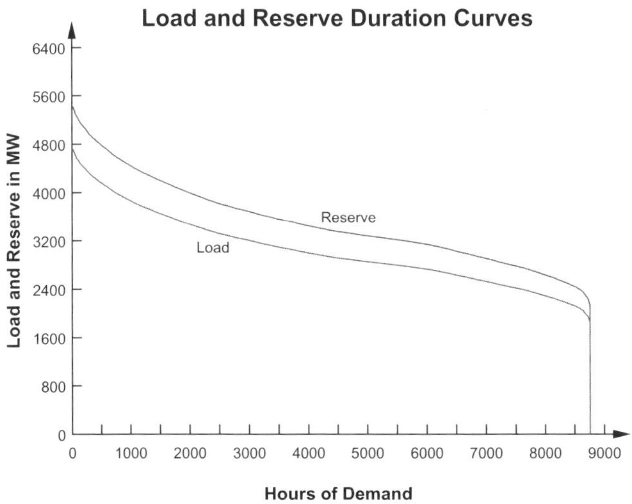

Figure 5.5

Table 5.2 Characteristics for Generation Expected to be On Line

Unit Name	Fuel	Max capacity	Heat rate	Fuel cost	VOM	Total variable cost
Neptune	Nuclear	1200	10.1	1.50	6.00	21.15
Venus I	Coal	900	10.0	2.80	4.50	32.50
Venus II	Coal	800	10.3	2.80	4.50	33.34
Orion	Gas	550	6.9	7.45	4.00	55.41
Pegasus	Gas	500	7.0	7.60	4.00	57.20
Mercury	Gas	250	10.5	7.45	4.15	82.38
Mars	Gas	250	10.7	7.45	4.15	83.87
Pluto	Gas	250	11.3	7.45	4.15	88.34
Haley	Gas	250	12.5	7.45	5.00	98.13
Xena	Gas	250	12.5	7.45	5.00	98.13
Kuiper	Gas	150	13.0	7.45	5.00	101.85

Table 5.3 Characteristics of Technologies Under Consideration

Technology type	Fuel	Max capacity	Heat rate	Fuel cost	VOM	Total variable cost
Baseload	Coal	To be selected	8.5	2.8	5.00	28.80
Intermediate	Gas	To be selected	6.8	7.45	6.25	56.91
Peaker	Gas	To be selected	10.0	7.45	5.00	79.50

Additionally, financing terms such as the interest rate and time horizon of the loan are considered. From this the total annual payment to finance construction is given as follows:

$$\text{Cost of Construction} = \sum_{j=1}^{T} \frac{\text{AP}}{(1+\text{interest})^j} = \text{AP}\sum_{j=1}^{T} \frac{1}{(1+\text{interest})^j}$$

$$\text{AP} = \frac{\text{Cost of Construction}}{\sum_{j=1}^{T} \frac{1}{(1+\text{interest})^j}}$$

- T is the time period of the loan
- AP is the total per MW annual payment

As an example, consider the annual payment for a CCGT. From Section 2.3.8 it is seen that a CCGT costs \$850,000 per MW. Assume a 30-year loan at 9 percent interest. This gives the per MW annual payment as the following:

$$\text{AP} = \frac{850{,}000}{\sum_{j=1}^{T} \frac{1}{(1+.09)^j}}$$
$$= \$82{,}736/\text{MW}$$

Table 5.4 represents the fixed cost for each of the three technologies.

Table 5.4 Fixed Costs of Technology Under Consideration

Unit type	Fixed cost, \$/MW
Baseload	155,738
Intermediate	82,736
Peaker	63,270

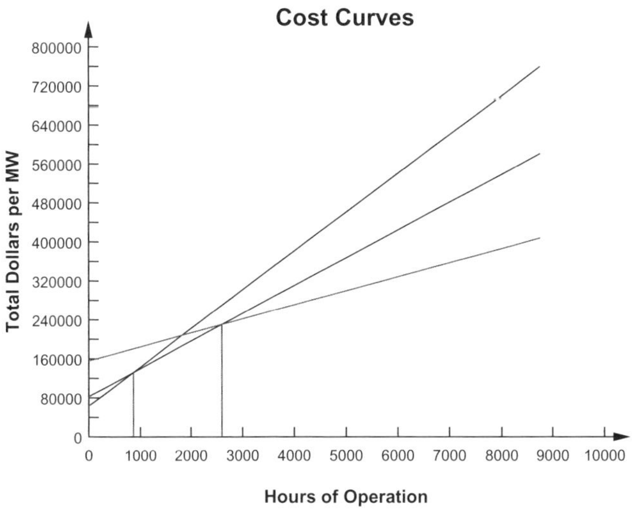

Figure 5.6

Step 4. Determine the dispatch levels of the break points where the optimal choice among the technology selections transitions from one technology to another.

With the fixed and variable costs, the cost curves for the three technologies are as follows:

$$\text{CostB}(X) = 155{,}738 + 28.80X$$

$$\text{CostI}(X) = 82{,}736 + 56.91X$$

$$\text{CostP}(X) = 63{,}270 + 79.50X$$

- X is the hours of operation
- The units for cost curves are \$ per MW

Figure 5.6 provides a graph of these curves. The relevant points of intersection are $X = 2597$ hours and $X = 862$ hours. Above 2597 hours of operation, baseload technology is optimal. Between 862 and 2597 hours of operation, intermediate technology is optimal. Less than 862 hours of operation, peaker technology is optimal.

Step 5. Develop a merit order stack for all generation; this includes current generation, planned generation, and technologies that are under consideration for selection.

Table 5.5 provides a capacity stack for year 5. Note that the capacity of the baseload technology is set at zero because there is insufficient development time. The capacity for year 6 depends on the selections made in year 5.

Table 5.5 Merit Order Stacks

Unit name	Fuel	Max capacity, MW	Total variable cost, $/MWH
Neptune	Nuclear	1200	21.15
Baseload	Coal	0	28.80
Venus I	Coal	900	32.50
Venus II	Coal	800	33.34
Orion	Gas	550	55.41
Intermediate	Gas	To be determined	56.91
Pegasus	Gas	500	57.20
Peaker	Gas	To be determined	79.50
Mercury	Gas	250	82.38
Mars	Gas	250	83.87
Pluto	Gas	250	88.34
Haley	Gas	150	98.13
Xena	Gas	250	98.13
Kuiper	Gas	150	101.85

Step 6. Stack the capacity of the current and planned generation units as well as the selection technologies into the load duration curve in merit order. The MW volumes for the selection choices is determined by the minimum of the volume that places the next technology selection at its break point or the volume that fills the total annual MW requirement including reserve.

For year 5, we first note that the total amount of current and planned capacity is 5300 MW and the requirement for load and reserves is 5520 MW. The total required capacity additions amount to 220 MW. For year 5, there is 3900 MW of expected capacity that is more efficient than the peaker technology, Neptune through Pegasus. From Step 5, the peaker technology is optimal for load less than 862 hours. Examining year 5's load duration curve, the MW band associated with the 862-hour point is 3976 MW. This means that an additional 76 MW of intermediate capacity is required to ensure that any additional peaker technology operates less than 862 hours. The remaining capacity need, 144 MW, is filled with peaker technology. Figure 5.7 provides a graph illustrating the capacity additions. The intermediate technology fills the shaded area between Orion and Pegasus, while the peaker fills the area between Pegasus and Mercury. Note that the peaker technology fills a band that operates at less than 862 hours, while the intermediate technology fills a band that operates above 862 hours.

Step 7. By defining the incremental load as the load that the selection technologies fill on being stacked under the load duration curve, the MW volumes selected in Step 6 provide the solution to the optimization problem.

Define the incremental load as the load associated with the intermediate and peaker bands. The optimization problem is solved.

We do not solve the year 6 problem, but note that the selection choices of 76 MW of intermediate technology and 144 MW of peaker technology are included as planned technology in the year 6 stack. Then the same steps for the year 5 problem are carried out and executed. For the year 6 problem, there is the potential for selecting new baseload construction since a unit can come on line within this time frame.

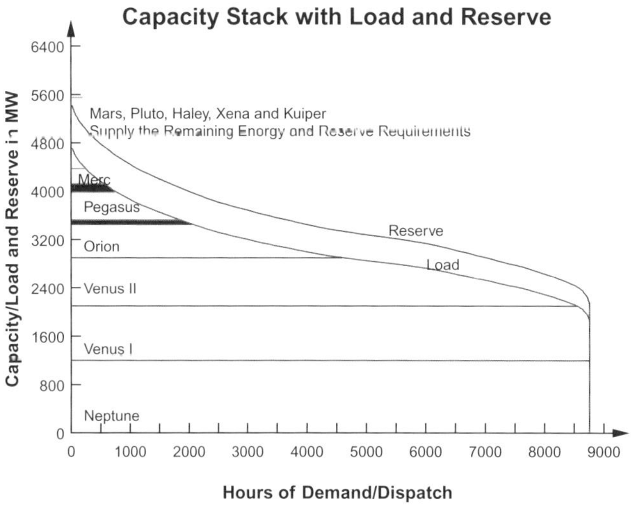

Figure 5.7

Remarks

- The salient feature of the example is that the solution is a trade-off between fixed and variable costs. The cost curves in Step 4 illustrate the trade-off. The solution fills the incremental load with the lowest-cost alternative among the technologies. The break points are points at which the technology choice changes from one technology to another. The trade-off between fixed and variable costs is a feature of all selection methods. The following sections approach this trade-off in a more sophisticated way.
- The costs associated with Step 4 require more thought than the simple calculation that is presented. The calculation assumes that a dollar in years 5 and 6 has the same value. This is generally not the case.

5.4.7 Improved Optimization Problem

One criticism of the above method is that the incremental load is defined in an artificial manner. This allows for a simple solution method, but places an artificial constraint on the problem; a solution with this artificial constraint is not as good as one without this constraint. Below, we pose a more robust formulation of the problem that leads to an improved solution.

5.4.7.1 In Words

- Objective: For a given year, minimize the total cost of capacity additions in MW by technology.
- Constraints:

1. Capacity constraints: There must be enough capacity additions to meet load and reserve requirements.
2. Development time constraints: Capacity may not come on line before the time required by the development cycle.

5.4.7.2 In Equations

- Objective

$$\underset{\mathrm{MW}_j^y}{\mathrm{Min}}\left(\sum_{i=1}^{N}\mathrm{Variable}_i^y+\sum_{j=1}^{P}\mathrm{MW}_j^y(\mathrm{Fixed}_j^y+\mathrm{Variable}_j^y)\right)$$

  - The superscript y represents the year of interest and begins with $y = 0$ as the current year. This is only an identifier and does not indicate the mathematical operation exponentiation.
  - MW_j^y represents the capacity additions in MW of technology j for year y. These are the decision variables.
  - $\mathrm{Variable}_i^y$ represents the variable cost of unit i in year y. Units associated with the subscript i are assumed to be in operation by year y as opposed to the decision variables. The variable cost of units in operation depends on the decision choice, as any new additions affect the dispatch.
  - Fixed_j^y and $\mathrm{Variable}_j^y$ represent the fixed and variable costs of technology j during year y.
  - There are P technology types.
- Constraints

1. $\sum_{i=1}^{N}\mathrm{MW}_i^y+\sum_{j=1}^{P}\mathrm{MW}_j^y \geq \mathrm{MW}^y$
   - MW_j^y represents current and planned generation.
   - MW_j^y represents the capacity associated with the decision variables. There are P types of technology.
   - MW^y represents the capacity requirement for year y.
2. If $y < \mathrm{DT}_j$, then $\mathrm{MW}_j^y = 0$.
   - DT_j represents the development time of technology j.

The difference between this formulation and the one above is that total load requirements, rather than incremental load, are used to set the capacity constraint. The dispatch of operational capacity is considered in the optimization problem to minimize overall costs. This provides more flexibility to allocate capacity quantities among the available technologies, leading to an improved solution.

Remarks

- The solution demonstrates that the decision for technology choice is a balance between higher fixed costs with lower variable costs and lower fixed costs with higher variable costs. If there is a low energy output requirement, the technology with low fixed costs is preferred. As output energy requirements increase beyond certain thresholds, the preferred technology shifts to that with higher fixed costs but lower variable costs.

- The study above is handicapped because it makes the optimal selection choice over a single year's time horizon. The lifetime of generation units is rarely less than 25 years and in many instances significantly beyond that. The trade-off between fixed and variable costs should be addressed over a longer time horizon.
- The final selection choices provide only guidance. Actual recommendations for specific projects must be made in coordination with project developers and transmission analysts as discussed in Section 5.1.
- One drawback that is typical of planning studies is that the recommended quantities by year are not necessarily commensurate with actual unit specifications. The above example illustrates this drawback. In this example, the recommended addition of CCGT capacity for year 5 is 76 MW. However, CCGT units are much larger in scale, and construction of a 76 MW unit is not realistic. Nevertheless, a multiyear study provides guidance.

5.5 GENERATION ADDITIONS AND RETIREMENTS WITHIN A SINGLE CONTROL AREA

In this section we present another formulation of the optimization problem. This formulation is much more complex than that of the preceding section. The approach integrates selections over a multiyear study into a single optimization problem as opposed to making a separate selection each year. Also, capacity retirements are considered along with capacity additions. The possibility of incorporating a more realistic dispatch as presented in Section 4.6 is also discussed. Because of these complexities, a solution method is beyond the scope of this text. However, some issues concerning a solution method are discussed.

The problem is formulated in line with Section 4.6. There, optimal dispatch in a single control area is presented. Consideration of the problem within a single control area is not overly restrictive. Although there are exceptions, generation is normally located in proximity to demand centers. This is necessary for providing reactive power. Additionally, it is generally more economical to locate generation close to load centers than to locate generation at a distance and build transmission. This is due to two factors, the cost of construction for transmission and power losses that occur along transmission paths. Nevertheless, there are circumstances under which distant generation resources provide significant power to a given locality. These circumstances are discussed in Section 5.6 below.

Returning to the concept of most economical build and retirement decisions, as with the previous section there are two cost contributions that must be considered, fixed and variable. For our purposes all costs in the economic dispatch decision of Chapter 4 are variable costs, while all other costs are fixed. The goal is to minimize the present value of the overall costs over the time horizon of the forecast, typically on the order of 20 years. To determine the variable costs one must perform an economic dispatch as described in Chapter 4 over the entire time horizon of the study. Then the fixed costs are added, and the algorithm must make optimal build and retirement decisions over the combined fixed and variable costs.

There are two points of interest concerning this approach. The first is that the structure requires an optimization problem within an optimization problem. The interior optimization problem is the economic dispatch from Chapter 4; an economic dispatch must be solved for over the entire study horizon. The exterior optimization problem is a selection of generation additions and retirements that minimizes the present value of overall costs.

The second point of interest is a consequence of the first. Setting an economic dispatch over the entire time horizon requires considerable computational time. The time requirement is necessarily magnified since the economic dispatch will have to be done over many different possible generation portfolios until the external optimization problem converges to an optimal solution. The time expense must be reduced through a combination of methods. Two possibilities are mentioned below.

- Select representative time periods for performing the economic dispatch rather than dispatching over every single hour of every day. Some planning departments use a typical week from each month. Others select typical weeks by season.
- Do not dispatch at the hourly level. Instead, consider time blocks corresponding to demand patterns. The time blocks should have the resolution to separate peaking units (combustion turbines) from intermediate units (combined cycles). Six 4-hourly time blocks are sufficient.

The remainder of this section formulates the optimization problem. The presentation follows our standard: First assumptions are set, and then the formulation is presented in words and equations.

5.5.1 Assumptions

The following is a list of assumptions.

- There is a list of technology possibilities for new construction. For each technology this list specifies all costs and operational parameters over the lifetime of the project. For each technology the solution provides the number of MW of new construction that should be built. Technology choices may be CCGTs, CTs, coal-fired steam turbines, and others.
- A completion time for each technology is specified. The completion time includes all permitting, purchasing of supplies, construction, and testing.
- Fixed costs are given as annual figures normalized by MW for new construction.

- It is possible to disaggregate the environmental costs into fixed and variable cost components. The fixed cost is the capital cost of emissions pollution control devices. A variable cost is available by technology and fuel. This is a multiplier on the quantity of fuel burned.
- A least-cost dispatch is set to determine variable costs. This section assumes a least-cost dispatch in accordance with Section 4.6. All assumptions in that section apply.
- Retirement costs are negligible. Except for retirement of nuclear facilities, this is reasonable.
- A value of remaining generation at the end of the study is available.
- Retirements and additions occur on an annual cycle at the same time. This allows one to state the problem in annual time periods. It is possible to state the problem in shorter time frames, but then the computational time increases.

5.5.2 In Words

- Objective

  The objective is to minimize the present value of all fixed and variable generation costs over the time horizon of the study. The decision variables available for attaining this objective are retirement of existing units and construction of new units.
- Constraints
  1. New construction cannot come on line in advance of its completion time.
  2. There must be sufficient resources to meet forecasted load, primary reserve, and secondary reserve requirements over the time horizon of the study.
  3. All constraints from Section 4.6, modified for the selected time blocks

5.5.3 In Equations

- Objective:

$$\min_{a_{jk}, b_{jkm}} \left[\sum_{j,k} (D^k a_{jk}(F_{jk} + V_{jk})) + \sum_{j,k,m} (D^k b_{jkm}(f_{jkm} + v_{jkm})) - \sum_{j,m} (D^N (\mathrm{EV} a_{jN} + \mathrm{EV} b_{jNm})) \right]$$

where

- D^k is the discount applied to year k.
- $a_{jk} = 1$ if unit j remains in operation during year k, $= 0$ if unit j has been retired during year k or before year k.

- b_{jkm} = the capacity of technology j in year k that came on line in year m.
- N = final year of the investigation.
- $\mathrm{EV}a_{jN}$ = remaining value of unit j at the end of year N.
- $\mathrm{EV}b_{jNm}$ = remaining value of technology j in year N that was constructed during year m.
- F_{jk} = the fixed cost of unit j in year k.
- V_{jk} = the variable cost of unit j in year k as determined by an economic dispatch.
- f_{jkm} = the normalized fixed cost of technology j that came on line in year m during year k, normalized by MW in year 1, \$/MW.
- v_{jkm} = the normalized variable costs as determined by an economic dispatch of technology j that came on line in year m during year k, normalized by MW online in year m, \$/MW.

- Constraints:
  1. $b_{jkm} = 0$ for all $k < m$. $b_{jkm} = b_{jkk}$ for all $k \geq m$.
  2. For all k, $\sum_{j,m} a_{jk}\mathrm{MW}_j + b_{jkm} \geq \mathrm{Max}L_k + \mathrm{RReq}_k$
     - MW_j is the capacity of unit j.
     - $\mathrm{Max}L_k$ is the forecast of the annual peak load for year k.
     - RReq_k are the primary and secondary forecasted reserve requirements for year k.
  3. All constraints from Section 4.6, modified for the selected time blocks

We end this section with several remarks.

Remarks

- The variable costs have not been explicitly written as equations. For each year, the variable costs of each unit are obtained by determining the economic dispatch of the unit over a sample period and then rescaling the results to match annual costs. To perform this, the economic dispatch problem of Section 4.6 must be solved for the sample period. Constraint Set 3 is used when solving the economic dispatch problem.
- The new construction by technology is explicitly expressed in terms of the year that the technology can come on line, the index m. There are several reasons for this feature. First, performance degradation can be applied in accordance with the number of years that a unit is on line. Second, performance improvements for succeeding generations of a specified generation technology may be considered. Third, the remaining value of the project can be more accurately represented.
- The value of generation at the end of the time period may be neglected if the time frame of the study is sufficiently long.
- The dispatch cost is the same as the variable cost. While the simulated dispatch technique is that of Section 4.6, the same formulation could apply to another simulated dispatch technique with modifications to Constraint Set 3. For example, one could use the merit order dispatch of Section 4.3, which is

once again used in Section 5.4. This would greatly simplify computation at the expense of accuracy of the dispatch. The difference between this approach and that of Section 5.4 is that this approach allows for multiyear optimization. Also note that the statement of the problem remains unchanged regardless of the dispatch choice, except for Constraint 3. This constraint would be modified to match the dispatch simulation method.

5.6 GENERATION ADDITIONS AND RETIREMENTS WITH TRANSMISSION TO A SINGLE CONTROL AREA

This section builds on Section 5.5 by incorporating the potential for construction of plants outside the control area and incurring additional transmission expenses. As stated in Section 5.5, there are special situations in which this is an economical alternative to construction within the control area. One special circumstance is the construction of large-scale hydroelectric facilities. Facilities such as Hoover Dam and systems along the Columbia River in the Pacific Northwest produce surplus power beyond local requirements. It is, however, economical to build such systems and transmit power to distant control areas. Both the Hoover Dam and Northwest systems have transmission capacity into Los Angeles. In the US there are few potential remaining sites for large-scale hydroelectric projects. Accordingly, it is unlikely that there will be projects of this nature in the foreseeable future.

More likely, remote construction will be driven by environmental considerations. For example, EPA and state emissions standards are difficult to meet in the Los Angeles basin. This provides an incentive to construct new generation in remote locations.

Another driver of remote construction is land costs. In certain locations, land within a given control area has become sufficiently expensive to warrant remote construction of power generators. Additionally, future siting of nuclear units will most likely occur in remote areas.

The formulation of the optimization problem follows that of Section 5.5, with minor alteration. Indeed, the textual formulation is identical to that of Section 5.5 and is not repeated. For the mathematical formulation, all that is necessary is to increase and redefine the set of variables representing technologies. This set should represent both technology choices and locational choices. The fixed and variable costs must incorporate cost of construction for transmission as well as costs for delivery into the original control area.

More specifically, the optimization problem is defined as follows.

- Objective

$$\underset{a_{jk},b_{jkm}}{\text{Min}}\left[\sum_{j,k}(D^k a_{jk}(F_{jk}+V_{jk}))+\sum_{j,k,m,p}(D^k b_{jkmp}(f_{jkmp}+v_{jkmp}))\right.$$
$$\left.-\sum_{j,m}(D^N(\text{EV}a_{jN}+\text{EV}b_{jNm}))\right]$$

- D^k is the discount applied to year k.
- $a_{jk} = 1$ if unit j remains in operation during year k,
  $= 0$ if unit j has been retired during year k or before year k.
- b_{jkmp} = the capacity of technology j in year k that came on line in year m at location p.
- N = final year of the investigation.
- $\text{EV}a_{jN}$ = remaining value of technology j at the end of year N.
- $\text{Ev}b_{jNm}$ = remaining value of technology j in year N that was constructed during year m.
- F_{jk} = the fixed cost of unit j in year k.
- V_{jk} = the variable cost of unit j in year k as determined by an economic dispatch.
- f_{jkmp} = the normalized fixed cost of technology j that came on line in year m during year k at location p, normalized by MW in year 1, \$/MW.
- v_{jkmp} = the normalized variable costs as determined by an economic dispatch of technology j that came on line in year m during year k at location p, normalized by MW online in year m, \$/MW

- Constraints

  1. $b_{jkmp} = 0$ for all $k < m$; $b_{jkmp} = b_{jkkp}$ whenever $k \geq m$.

  2. For all k, $\sum_{j,m} a_{jk}\text{MW}_j + b_{jkmp} \geq \text{Max}L_k + \text{RReq}_k$

  - MW_j is the capacity of unit j.
  - $\text{Max}L_k$ is the forecast of the annual peak load for year k.
  - RReq_k are the primary and secondary forecasted reserve requirements for year k.

  3. All constraints from Section 4.6, modified for the selected time blocks

Note that the difference between this formulation and that of Section 5.5 is the introduction of an additional index, p. This index indicates location. Locational differences are then set for fixed and variable technology additions and include transmission expenses.

5.7 GENERATION ADDITIONS AND RETIREMENTS AND TRANSMISSION ADDITIONS WITHIN A NETWORK

This section presents an integrated formulation of both generation and transmission choices within the context of a network. These studies can address additional considerations that the previous study cannot address. For example, seasonal consumption patterns differ between different locations, providing possibilities for the long-term allocation of new construction. The Pacific Northwest consumes more power in the winter than the summer, whereas the Southwest consumes more power in the summer than the winter. The analysis formulated in this section would provide a basis for cost sharing and allocation of new construction between counterparties in the Northwest and Southwest.

The approach is very similar to that of Section 5.6, the single control area case. The single control area case is based on minimization of fixed and variable costs asso ciated with dispatch in a single control area, and variable costs are calculated with the corresponding economic dispatch in Section 4.6. By analogy, the formulation in a network is also based on minimization of fixed and variable costs across the network, including construction costs. Additionally, the variable costs are calculated with the corresponding economic dispatch in Section 4.7. Because of the size of the problem it is not feasible to perform the power flow analysis, the Phase 2 component of the economic dispatch described in Section 4.7. The formulation is presented below.

5.7.1 Assumptions

All the assumptions from Section 5.5 apply. In addition, there is an assumption that there are no transmission line retirements.

5.7.2 In Words

- Objective

  The objective is to minimize the net present value of all fixed and variable generation costs throughout the network. The decision variables available for attaining this objective are retirement of existing units, construction of new units, and transmission lines.

- Constraints

  1. New construction cannot come on line in advance of its completion time.
  2. There must be sufficient resources to meet forecasted load, primary reserve, and secondary reserve requirements over the time horizon of the study for every control area. Sufficiency may be met through imports provided that generation is only allocated to a single control area. Reserve requirements for transmission must also be met.
  3. All constraints from Section 4.7, modified for the selected time blocks

5.7.3 In Equations

- Objective

$$\min_{a_{jkp},b_{jkmp},c_{jkmp}} \left[\sum_{j,k} (D^k a_{jkp}(F_{jkp}+V_{jkp})) + \sum_{j,k,m,p} (D^k b_{jkmp}(f_{jkmp}+v_{jkmp})) + \sum_{j,k,m,p} (D^k c_{jkmp}(t_{jkmp}+w_{jkmp})) - \sum_{j,m,p} (D^N(\mathrm{EV}a_{jNp}+\mathrm{EV}b_{jNmp})) \right]$$

- D^k is the discount applied to year k.
- $a_{jkp} = 1$ if unit j in area p remains in operation during year k, $= 0$ if unit j has been retired during year k or before year k.
- b_{jkmp} = the capacity of technology j in year k that came on line in year m and in area p.
- N = final year of the investigation.
- $\mathrm{EV}a_{jNp}$ = remaining value of unit j in area p at the end of year N.
- $\mathrm{Ev}b_{jNmp}$ = remaining value of technology j the end of year N that was constructed during year m in region p.
- F_{jkp} = the fixed cost of unit j in year k in area p.
- V_{jkp} = the variable cost of unit j in year k in area p as determined by an economic dispatch.
- f_{jkmp} = the normalized fixed cost of technology j that came on line in year m in area p during year k, normalized by MW in year 1, \$/MW.
- v_{jkmp} = the normalized variable costs as determined by an economic dispatch of technology j that came on line in year m in region p during year k, normalized by MW online in year 1, \$/MW.
- c_{mpq} = the total additional transmission capacity from region p to region q that is added to the network in year m.
- t_{kmpq} = the fixed cost of transmission capacity from region p to region q that is added to the network in year m during year k.
- w_{kmpq} = the variable cost of transmission capacity from region p to region q that is added to the network in year m during year k.

• Constraints

1. $b_{jkmp} = 0$ for all $k < m$; $b_{jkmp} = b_{jkkp}$ for all $k \geq m$.

2. For all kp, $\sum_{j,m} a_{jkp}\mathrm{MW}_{jp} + b_{jkmp} \geq \mathrm{Max}L_{kp} + \mathrm{RReq}_{kp}$

- MW_{jp} is the capacity of unit j in node p.
- $\mathrm{Max}L_{kp}$ is the forecast of the annual peak load for year k in node p.
- RReq_{kp} are the primary and secondary forecasted reserve requirements for year k in node p.

3. All constraints from Section 4.7, modified for the selected time blocks

Remarks

- Note that the power flow problem, Phase 2 of Section 4.7, is not addressed. There is too great a computational time expense to include this in the analysis. Instead, the output from this problem should be used to perform a separate power flow analysis that can then be used to develop transmission plans. Aside from a power flow analysis, other studies must be performed as part of the development of transmission planning. These include studies that determine system stability in the event of equipment failure or other perturbations to the system, e.g., lightning strike.
- Transmission costs for system upgrades associated with new construction have not been addressed in the above formulations. As an estimate, one could associate these costs with the fixed cost of construction.

- The second constraint requires that a feasible solution across the network would allow each control area to meet its load and reserve requirements. This solution has nothing to do with actual daily energy flows; it only demonstrates the potential to meet load and satisfy reserve requirements.

5.8 RESERVE REQUIREMENTS

We close this chapter with a note on reserve requirements. A probabilistic approach is taken to this problem as reserves are established to meet contingencies and contingencies themselves are quantified as probabilistic events. A preliminary step for setting reserve requirements is to establish an acceptable level of unintended service interruptions. This is expressed as an acceptable probability of not being able to fully service load. For example, a utility might base its reserve margin for a given year on a calculation that says it must have a 99.5 percent chance of fully servicing its customer demand over the course of the year.

Once a target is established, the factors that determine the ability to service demand are identified. These include the following: full customer demand, the total capacity of the system, maintenance schedules, forced outage rates, capacity derates due to mechanical uncertainties, import capabilities, grid performance, and others.

The concept is to develop a functional relationship between the demand that is serviced and the factors identified above. This functional relationship is expressed as a mathematical model and provides a specific MWH quantity for each hour of the year. Finally, the factors are described probabilistically in a manner that allows for computer simulations of different outcomes of the factors. One performs the computer simulation of the outcomes on the load servicing factors, and, using the mathematical model that relates these factor outcomes to the quantity of load that is serviced, one arrives at a series of outcomes for the load that is serviced.

At this stage there is a series of outcomes of load that is serviced along with a corresponding series of total demand (load that is requested). Both series are given in MWH. Summing up the outcomes of load that is serviced and dividing by the sum of outcomes of total demand provides a ratio of serviced load over demand. This ratio must meet the target; as described above, the ratio must be greater than .995.

In the case that the ratio does not meet the target, additional capacity must be added to the system. The additional capacity is expressed in MW and is used in the constraints of the optimization problems described in the previous sections. This approach provides for the introduction of probabilistic outcomes into the optimization problem.

Chapter 6

Midterm Utility Planning

Chapter 4, *Short-Term Utility Planning*, addresses economic dispatch with a set portfolio of resources. Chapter 5 on long-term planning presents methods for constructing an optimal portfolio of power plants and transmission resources. In between these decisions are other issues that must be addressed: maintenance scheduling for both generation and transmission, use of energy-limited resources, and capacity contracting with other utilities. These issues constitute the realm of midterm planning. Although they affect short-term planning, they are outside the scope of short-term planning because these decisions are made long before the day-ahead schedule is set. Midterm planning is distinct from long-term planning because midterm planning addresses management of an existing portfolio, whereas long-term planning addresses the development of a future portfolio.

Section 6.1 provides the informational requirements for setting midterm plans. Section 6.2 concludes the chapter with the formulation of an optimization problem that establishes a midterm plan. Two examples of midterm planning are also provided.

6.1 INFORMATIONAL REQUIREMENTS

Midterm planning optimizes dispatch costs over a time horizon ranging between 1 month and 3 years. A complete description of the load and supply outlook is necessary. This section presents informational requirements that are specific to midterm planning addressing load, unit outages, energy-limited resources, and interutility contracts.

6.1.1 Load

As with short-term and long-term planning, load is the starting point for midterm planning because plans are set to supply load. Load can be forecasted but is uncertain

---

Electric Power Planning for Regulated and Deregulated Markets. By Arthur Mazer

over the planning horizon. Midterm plans must be set to meet many possible outcomes for the load, so planners must develop more than a single load forecast. There are different approaches to the problem.

One approach is to develop scenario outcomes. Three scenarios are frequently developed, a high case, a base case, and a low case. Each case uses different weather and load growth assumptions.

It is also common to adopt a probabilistic approach. With a probabilistic approach, many potential load outcomes may be developed and probabilities may be assigned to the outcomes.

Remark

- The normal distribution is an appropriate distribution for describing load outcomes. One can forecast the expected load with methods similar to those described for long-term planning. Historical information provides the standard deviation necessary to complete the description of the normal distribution.

6.1.2 Maintenance Outages, Generation and Transmission

Chapter 2 presents the various generation technologies and discusses their need for scheduled maintenance. Maintenance schedules must be set in accordance with personnel availability. Utilities do not have the in-house capability to perform all types of maintenance requirements. Typically, there are maintenance contracts between utilities and service providers along with a well-defined set of responsibility sharing. To perform maintenance, maintenance crews must be scheduled in advance. Arrangements for both in-house personnel and service provider personnel must be coordinated. Some maintenance procedures require the coordination of personnel up to a year in advance, with limited ability to change the schedule. For example, this is almost always the case with scheduled nuclear outages and retrofitting of turbines. Maintenance procedures that are completely managed in house may allow for some flexibility in scheduling personnel.

Aside from personnel requirements there are also contractual obligations that restrict the scheduling of maintenance outages. Utilities purchase generation equipment along with equipment warranties. Turbines, for example, are typically under warranty. Manufacturers' warranties specify maintenance scheduling requirements, and maintenance must be performed within the specification or the warranties become invalid. Transmission equipment also requires scheduled maintenance and upgrades. As with generation maintenance, utilities must assemble equipment and personnel to perform maintenance and upgrades.

For all generation and transmission equipment, maintenance scheduling requirements must be maintained within a data set. The requirements specify a time range within which the outages must take place and the duration of outages.

6.1.3 Forced Outages

Aside from planned outages, midterm planners must be aware of forced outages. Forced outages are normally described probabilistically. Expected forced outage rates and durations can be extracted from a historical data set. It is the possible to develop scenarios and probabilities for future unit availability.

6.1.4 Wind and Hydro Resources

Wind power is an example of a resource in which the energy output is determined by weather conditions. Other than maintenance scheduling, it is not possible to optimize wind production: The production decision is made by forces of nature. Nevertheless, an outlook of production is required to optimize other resources in the supply portfolio. For facilities that have been in place over a sufficiently long time frame, the production forecast can be set by using an agreed-upon scenario of historical production. Otherwise, forecasts must be developed with a wind forecast and models that provide production capability as a function of wind speed.

Hydroelectric generation is another technology in which the energy output is determined by weather. Unlike wind generation, for certain hydroelectric systems there may be opportunity to schedule the output by controlling water flow rates. Consider, for example, a mountain system having a network of reservoirs at different elevations, with the higher elevation reservoirs feeding the lower elevation reservoirs. The holding capacity of the reservoirs frequently allows for energy storage over several weeks and production made available when it is most needed. An interesting aspect of the problem is the interplay between reservoir levels at different elevations. Timing of the energy production by releasing water from the higher level reservoir affects production potential at the lower elevation reservoir. The production may be optimized to provide the most economical use of the hydroelectric resource.

Output is limited by the weather, configuration of the hydro network, and legal requirements that effect reservoir management. Accordingly, hydro resources are energy constrained because of both weather and regulatory factors. As energy is stored in hydro reservoirs that are filled by precipitation, the weather is the dominant factor that determines hydro power production. The configuration of the hydro network determines storage capacity and influences operational flexibility: the ability to time energy production. Legal requirements also influence operational flexibility. Legal requirements include requirements to maintain reservoir levels within certain bounds as well as additional requirements on flow rates.

Midterm planners must establish an initial data set that encompasses hydro production ability and constraints. A necessary input into the data set is a forecast by month of water flow rates entering the hydro systems. One can accomplish this in several ways. First, statistics on historical flow rates can be obtained and planners can generate a scenario from the history. Another approach is to predict flow rates with a weather scenario. Either a forecast of the weather or an agreed-upon scenario from climatic data can be used.

Environmentally-Constrained Energy-Limited Resources

Regulatory limitations on emissions also limit the energy availability of some plants. Production constraints are a result of local and federal environmental statutes that limit the quantity of NOx and SOx emissions. This is accomplished through a permitting process that authorizes a specified level of emissions. For each generation plant, the allowable emissions must be placed into a data set.

6.1.5 Interutility Contracts

Within a regulated environment, neighboring utilities enter into contracts to exchange capacity and energy as well as to provide access to grid services. A common transaction is a capacity transaction. A utility that does not have the required quantity of generation capacity available for meeting regulatory requirements may on a temporary basis lease generation capacity from another utility that has surplus capacity. The leasing utility typically pays a leasing fee to the utility that owns the capacity. In addition to the leasing of generation capacity, the transaction often includes transmission capacity and services that guarantee power transfer capability on pathways connecting the acquired generation facility with the leasing utility's control area. Through such transactions the utility requiring additional capacity is able to meet regulatory capacity requirements. Additionally, the utility that supplies the capacity is able to earn revenues through the transaction.

Capacity transactions occur when surplus capacity from one utility is available to fill the capacity deficit of another utility. Deficits may occur as a result of planned outages, extended unplanned outages, poor hydro conditions, or load growth that exceeds construction of new generation.

Seasonally varying capacity conditions may occur with seasonally varying hydroelectric production capability. During seasons in which hydroelectric production is high, surplus capacity may be sold to a neighboring utility. However, during low-production seasons, the utility owning the hydroelectric facility may require temporary capacity from a neighboring utility.

Capacity transactions also occur between utilities having complementary seasonal loads. Consider, for example, the case of a utility in an area with mild summers but cold winters having winter-peaking load and another utility in an area with hot summers that result in air conditioning-driven summertime peak loads. Capacity transactions would allow these utilities transfer capacity in accordance with needs.

A typical structure for a capacity transaction is one in which there is a fixed payment, called a capacity payment, and an energy payment. The capacity payment is a fixed amount that transfers capacity and dispatch rights to the purchasing utility. The energy payment is a dollar per MWH payment that the purchasing utility pays to the owning utility for every MWH that the generation unit underlying the agreement produces.

One aspect of midterm planning is support of decision making for capacity transactions. The first step is to design a data set that determines a capacity price and an associated energy price for interutility transactions.

6.2 FORMULATION OF THE OPTIMIZATION PROBLEM

Midterm planning is the decision to include or exclude a resource from a utility's capacity base in accordance with maintenance requirements, weather and environmental limitations, or the price and availability of capacity within a network of utilities. The previous sections provide the specifics of informational requirements for making decisions. This section formulates the optimization problem in which information is processed into decisions.

The long-term planning problem addressing capacity additions and retirements is also a decision to include or exclude a resource within the resource base. Accordingly, the formulation of the midterm optimization problem is similar to that of the long-term optimization problem. Indeed, the formulation given in this section is similar to that in a section of the chapter on long-term planning, Section 5.7.

Recall that the goal of long-term planning is to minimize the cost of servicing load by making the most prudent transmission and generation capacity additions and retirement decisions. In this section the problem is identical: The goal is to make midterm decisions that minimize the cost of providing a specified load. The formulation of the long-term optimization problem incorporates an economic dispatch to determine dispatch costs. Identically, the formulation of the midterm planning problem also incorporates an economic dispatch for the same purpose. Another similarity between the midterm problem and the long-term problem is that the overall time horizon is broken down into shorter time frames. In the case of the long-term problem, a 20- or 30-year study is broken down into annual time frames; decisions are made on an annualized basis. In the case of the midterm problem, a 2- or 3-year study is broken down into weekly or biweekly time frames, and decisions are made over these shorter time frames.

While there are similarities between the long-term and midterm optimization problems, there are also differences. Unlike the long-term problem, for the midterm the fixed costs associated with the utility-owned resources are set; they are not influenced by decision variables. For the utility-owned resources, it is solely the variable costs over which the optimization is performed. Another difference is the inclusion of a capacity market between utilities. As stated above, such markets do exist within traditional utility environments.

In presenting the formulation we follow the standard used throughout the text. First we present the assumptions and then the formulation of the optimization problem in words and in equations. Recall that the long-term planning optimization problem is comprised of two components, a selection of resources and an underlying economic dispatch based on the resource selection. The midterm optimization is similar. Below we indicate the selection component of the problem; the dispatch is managed as described in the preceding chapters.

6.2.1 Assumptions

- It is possible to establish a utility market price of capacity and associated energy.
- The contracts and units that are available are dispatched economically to serve load as described in Chapter 4.
- It is possible to equate emissions output with MWH of production.

6.2.2 In Words

- Objective: The objective is to minimize the present value of the dispatch costs for serving load. The decision variable available for attaining this objective is the scheduling of the availability of each resource and capacity purchases and sales.
- Constraints:
  1. Unit maintenance constraints and refueling constraints
  2. Transmission maintenance constraints
  3. Emissions constraints
  4. Hydro constraints
  5. There must be sufficient resources to meet forecasted load, primary reserve, and secondary reserve requirements over the time horizon of the study. Sufficiency may be met through imports provided that generation is only allocated to a single control area. Reserve requirements for transmission must also be met.
  6. All constraints from Section 4.7, modified for the selected time blocks

6.2.3 In Equations

- Objective:

$$\underset{a_{jkp},b_{kpq},c_k}{\text{Min}} \sum_{j,k,p} D^k (a_{jkp}(V_{jkp}) + c_k M_k)$$

  - D^k is the discount applied to time period k.
  - a_{jkp} = 1 if unit j in area p remains in operation during week k or is under contract from a neighboring utility,
    = 0 if unit j is assigned maintenance during week k.
  - V_{jkp} = the variable cost of unit j during week k in area p. This cost is dependent on the resource selection.
  - c_k is the volume of capacity that is under contract from neighboring utilities or that is being contracted out to neighboring utilities. A positive value for c_k indicates that the capacity is under contract from a neighboring utility; a negative value indicates that capacity is contracted out to a neighboring utility.

- $b_{kpq} = 1$ if the transmission line between buses p and q is in service during week k.
  Otherwise, $b_{kpq} = 0$.
  The variables b_{kpq} are not explicitly expressed in the objective function but lie in the constraint set that determines the underlying dispatch.

- Constraints:

1. $\sum_{k=s}^{t} 1 - a_{jkp} < m_{jp}$
   - s is the first week that an outage can be scheduled.
   - t is the last week that the unit can be out.
   - $m \leq t - s$, and m represents the number of required maintenance weeks for unit j in area p.

2. $\sum_{k=s}^{t} 1 - b_{kpq} < tm_{pq}$
   - s is the first week that a transmission outage can be scheduled.
   - t is the last week that a transmission line can be out.
   - $tm_{pq} \leq t - s$, and tm_{pq} represents the number of required maintenance weeks for the transmission line running between points p and q.

3. $\sum_{k=s}^{t} \text{MWH}_{jkp} < \text{Max}_{jkp}$
   - s and t are bounds for emission constraint definitions. Usually emission constraints are given as an annual limit, so s represents the first week of a year in the study and t represents the last week of the same year.
   - MWH_{jkp} is the energy production of unit j in location p during week k. This is obtained from the underlying dispatch.
   - Max_{jkp} is the maximum allowable energy production as determined by emission constraints.

Constraints 4 through 6 are placed directly into the underlying economic dispatch problem. These are the constraints from the optimization problem presented in Section 4.7.

Remarks

- The constraint associated with the hydro dispatch, Constraint 10 from the optimization problem in Section 4.7, is a dynamic constraint. The constraint changes as hydro conditions changes, reflecting dispatch decisions as well as natural water flow rates.
- Constraints associated with transmission path outages are placed into the network dispatch: Constraint 3 from the optimization problem in Section 4.7.

6.2.4 Scheduling the Refueling of a Nuclear Unit

As an example of the use of the optimization problem, consider the scheduling of refueling for a nuclear unit; fuel depletion may be controlled by production adjustments. Suppose that it is currently October and under normal operating conditions, baseload at maximum output, the next refueling scheduled for a 500-MW nuclear unit is due next August. The utility requires 400 MW of this capacity to meet its forecasted load and reserve requirements. The utility can decrement output by 54 MW and delay the refueling by 1 month. Alternatively, the utility can lease the required 400 MW of capacity from a neighboring utility during August. We assume there is spare capacity available at a price. Which option should the utility choose?

The solution is to run the optimization problem for each case. Total energy costs and capacity payments are compared. The alternative with the cheapest overall cost is selected.

6.2.5 Dispatch of Environmentally Constrained Energy-Limited Resource

As another example of the use of the optimization problem, consider the problem of determining the dispatch of an environmentally constrained unit. Dispatching a unit reduces the availability of the unit in the future. The decision to schedule a unit in the day-ahead plan rests on the value of including the dispatch in the day-ahead plan versus the value of availability in the future. A decision criterion for making this assessment must be passed on to short-term planners so that they can decide whether or not to schedule the unit.

One criterion that can be passed on is a cost adder into the VOM of the unit. The cost adder reflects the value of future availability, and the unit is not dispatched unless system marginal production costs exceed the total cost of the unit with the adder included. In other words, when system production costs attain a level that exceeds the production cost of the environmentally constrained unit and production costs include the value of saving the unit for future availability, then the unit is dispatched. Otherwise, the unit is not dispatched.

To determine the cost adder, the optimization problem can be executed over the remainder of the year with an initial estimate for the VOM cost adder. The value of the VOM cost adder is then adjusted upward if the simulated dispatch indicates that the unit's emissions exceed its allowable limit or downward if the simulated dispatch indicates that the unit's emissions are below its allowable limit. The final value for the cost adder is achieved when the simulated dispatch indicates that the unit's emissions are equal to the allowable emissions or when the cost adder is zero. In the case when the cost adder is zero, the unit is not environmentally constrained.

The value of the cost adder changes throughout the year and must be periodically recalculated.

We close this chapter with some remarks.

Remarks

- While cooptimization of all midterm decisions is desirable, existing software systems cannot achieve this. Different software is required for different components. For example, specialized software that optimizes the dispatch of hydro units is separate from packages that dispatch other technologies. Output from one study is used as input into another.
- All midterm plans should be subject to a power flow analysis to ensure that power flows resulting from the plan are feasible. This is accomplished by testing the plan against predefined scenarios after the plan has been set as opposed to performing the power flow analysis within the dispatch routine.
- Midterm planning is an ongoing activity. Adjustments are made as system conditions change.
- Probabilistic analysis is often used to address uncertainty. For example, the cost adder applied to an environmentally constrained unit may be calculated under multiple scenarios with probabilities assigned to each outcome.

Chapter 7

A Market Environment

Chapters 8, 9, and 10 present market participant interactions in a market environment. The presentation parallels Chapters 4, 5, and 6: short-term planning, long-term planning, and midterm planning. Before proceeding, it is necessary to describe a market environment so that market participant behavior can be placed in context; this chapter provides a description. The chapter emphasizes the need for a market to address grid operations and economic dispatch as presented in Chapters 3 and 4.

The market described here is a hypothetical market; although it is similar to PJM, there are differences. The purpose of presenting this hypothetical market is to simplify the exposition while maintaining the basic principles that allow for effective operations. These principles are first established. Then we present the market structure over different time frames: short term, long term, and midterm.

We concentrate our description on electric energy and capacity markets. This simplification ignores two related and important markets: fuels and transmission. The intention is not to diminish the importance of fuels and transmission; we acknowledge that these markets all impact on each other. However, by concentrating on electric energy and capacity markets, we can align market structure and decisions with the operations of power delivery that are discussed in previous chapters.

The chapter is organized as follows. Section 7.1 presents underlying principles of market design. Sections 7.2, 7.3, and 7.4 then address short-term, long-term, and midterm market structures.

7.1 PRINCIPLES AND ARCHITECTURE

There are two issues that market design must address: reliability and competitive retail choice. Market designers must balance these issues because, as we describe later, they are often not compatible. When these interests conflict, market designers often give

Electric Power Planning for Regulated and Deregulated Markets. By Arthur Mazer

preference to reliability. Indeed, in the US the reliability standards for markets with retail choice match the reliability standards of traditionally regulated markets.

It is possible to classify reliability concerns into two related categories, networking issues and supply. As Chapter 3 explains, networking involves the operations of all transmission system components as well as primary and secondary reserves to maintain system reliability. The operations include maintaining and dispatching ancillary services as well as voltage support and frequency maintenance. By its fundamental nature, from an operational perspective, it is impossible to have more than one agent operating the grid. More than one agent could not coordinate the setting of feasible schedules before delivery, and it is not possible to split the dispatch of ancillary services in real time among several agents. Indeed, one agent's actions might act against another agent's actions. As discussed below, from a market perspective, a single grid coordinator is necessary to allow all market participants free access to the grid. Accordingly, grid operations are the responsibility of a single regulated agent. This agent is called the Independent System Operator (ISO). The ISO is a public, nonprofit agency.

The unanswered question from the preceding paragraph is, "Why can't there be private ownership of transmission lines with owners operating their transmission services competitively?" In fact, transmission lines and equipment are frequently privately owned by utilities. The utilities must turn this equipment over to the ISO for daily operations and scheduling of power flows. If the owning utility did not turn its facilities over to the ISO, the utility could abuse its scheduling authority and not allow other market participants to use the transmission and networking services. Competing power producers could not deliver their power to customers.

This explanation leads to another question: "Why wouldn't other market participants build and operate more transmission networking systems so that they could deliver their own power?" Chapter 5 describes the development process for building transmission lines. It is a long, expensive process to build full-scale networks with wide reach. Transmission systems have been built by incumbent utilities and grown for over a century. Alternatives are not feasible from an economic or regulatory perspective. Additionally, even if overlapping transmission systems did exist, they would have to be interconnected. The reliability of the system would require a single agent to operate it.

As Chapter 4 shows, it is not possible to decouple the operations of the transmission system from the scheduling and dispatching of generation. Accordingly, this function falls within the purview of the ISO.

Although independent transmission networks are not possible, it is possible to develop independent power producers (IPPs) as well as independent load retailers. The independent power producer owns and operates generation units and sells their output in wholesale markets. As a market participant, an IPP should become proficient at construction and operation of units so that their all in production costs are competitive.

Load retailers procure power in wholesale markets on behalf of their aggregate retail customer base. A load retailer should become proficient at predicting its customer base's load requirements and procuring those requirements at competitive prices.

The architecture of power markets is that of decoupled production (IPP) and retailer companies competing to sell their services. This provides customers with the possibility of selecting their retail service provider from among several competitors. The physical positions of production and retail agents are balanced through an ISO that operates the transmission network. The financial positions of production and retail agents are cleared through wholesale markets. These mechanisms are discussed in Sections 7.2 and 7.4.

In addition to IPPs and retailers, there are integrated energy companies that perform the functions of both IPPs and retailers. A utility is an example of an integrated company. Aside from utilities, there are merchant energy companies that are integrated. Integrated energy companies also balance their positions through the ISO.

This is not a deregulated architecture. The ISO is a completely regulated entity, and there must be strict protocols for interaction between the ISO, IPPs, and retailers. Indeed, fair access to the grid is necessary to maintain viable markets. Additionally, regulators want to ensure the same level of reliability in the competitive markets as in traditional markets. The entire regulatory framework for reliability remains intact in most markets. Moreover, regulations have been necessary to ensure that suppliers do not exercise market power. The result is that the body of regulatory legislation and policing agents for competitive markets is greater than that for traditionally regulated markets.

There are electricity markets throughout the world. In the United States the market with the longest history is PJM (Pennsylvania, New Jersey, Maryland). Other markets operate in Texas, New England, and New York. California opened a market before PJM, but the result was not favorable. The market closed, and California is preparing itself for a second try. Outside the United States, Scandinavia has operated a successful wholesale market since 1995. England and Australia operate markets. Several countries in Europe are at various stages in implementing electricity markets.

All the markets have the common feature that there is an independent ISO charged with operating the grid. Differences in the markets arise because there is no natural point of separation between grid and generation activities. Policymakers may task the ISO with broad responsibilities as the steward of reliability or have minimal reliability requirements and allow market mechanisms to determine reliability standards such as reserve margins. Tasking the ISO with broad responsibilities limits the component of the power delivery chain that is truly deregulated.

The charge that each retailer must pay to the ISO for grid services is the same and completely fixed under a regulated pricing mechanism. Tasking the ISO with more responsibilities means that a higher percentage of charges that retailers must pass on to their end customers is set and established in a regulatory environment, making it more difficult for retailers to differentiate their services from one another. On the other hand, experience has shown that unless the ISO assumes these broad authorities, reliability may be impaired. This is the tension between reliability and unfettered competition that is mentioned above.

7.2 SHORT-TERM MARKET DESIGN: DAY-AHEAD SCHEDULING THROUGH REAL-TIME DELIVERY

In this section we identify the roles and responsibilities of the market participants and ISO from day-ahead setting of the dispatch schedule to real-time delivery. This section focuses on physical transactions, transactions that result in physical power flows. The short-term networking and planning activities required to ensure power delivery in traditionally regulated markets are presented in Chapters 3 and 4. The presentation culminates in a list given in Section 4.1. All these activities are necessary in a market environment as well. Aside from those activities, additional activities are necessary to coordinate balancing between market players through the ISO. The activities follow.

Activity 1: Day-Ahead Load Forecasting and Demand Bidding

As in a traditionally regulated environment, day-ahead load forecasting must be performed so that resources to meet demand can be scheduled. Retailers are responsible for procuring on behalf of their customer base. However, the ISO is responsible for grid reliability. Both retailers and the ISO have a stake in the load forecast and accordingly perform their own forecasts. The retailers should determine their requirements in a day-ahead forecast and inform the ISO of their day-ahead procurement requirements. The ISO cannot force retailers to procure in accordance with its own forecast. However, the ISO uses its forecast to determine ancillary service (AS) requirements, Activity 2 below.

To account for consumer price response to market prices, retailers provide the ISO with a demand schedule that incorporates interruptible load along with market prices that trigger interruption. An example of a demand schedule for a single hour is given in Table 7.1.

Activity 2: Day-Ahead Setting of AS Requirements

There are several AS requirements: primary and secondary reserves, regulation up and down, reactive power, voltage support, and frequency regulation. The setting of

Table 7.1 Retailer's Bid Table

Hour ending 2 P.M.	
Price in \$/MWH	Demand in MWH
<80	3800
80 to <175	3700
175 to <250	3550
≥250	3350

requirements for these services depends on the load forecast. The function of these services is reliability. Since the ISO is tasked with reliability, it is the role of the ISO to determine the requirements. The load forecast is critical for determining these requirements. For this reason the ISO must produce its own forecast as described in Activity 1 above.

Activity 3: Day-Ahead Supply Bidding into Energy Markets

Just as the retailers provide the ISO with load requirements, the producers provide the ISO with the price and capabilities of their supply resources. Each unit is bid into the market with an associated cost. The cost includes all variable price components and a profit margin that the owner wishes to receive. The cost is presented as a function of output along with a separate start-up cost. In addition to providing the costs, retailers also provide the operational parameters of each unit that are necessary to set an economic dispatch, as described in Section 3.7. These parameters include ramping rates, minimum downtime, minimum uptime, minimum output, and maximum output.

Activity 4: Day-Ahead Supply Bidding into Ancillary Service Markets

In addition to bidding energy prices into the day-ahead energy market, producers also bid into a market for each of the ancillary services. It is not possible to create competitive conditions for all ancillary services. For example, there are many instances in which a specific unit is required to provide voltage support in order to maintain grid reliability. The owner of a unit that is required to operate to support the transmission system has monopoly power; there is not a competitive bidding process. For such services there must be a regulated rate regime.

Services for which there may be competitive markets include primary and secondary reserves and regulation services. Bidders provide two prices into these ancillary service markets. First, they bid a price for making necessary capacity available. Second, they bid an energy price similar to the energy price bid into the day-ahead energy-only market, Activity 3. The energy prices are used to determine real-time dispatch of ancillary services, as described in Activity 7.

Bidders into the ancillary service market must have equipment to provide the services for which they are bidding. For example, regulation requires the availability of AGC.

Activity 5: Determining the Economic Dispatch, Hourly Price, and AS Suppliers

Using the information provided in the supply and demand bids, the ISO performs an economic dispatch as described in Section 3.7. Recall that this dispatch is set within a network and includes a power flow analysis to ensure the feasibility of the solution. The economic dispatch identifies generators that are called on for energy and ancillary services. From this dispatch the ISO sets a unit production schedule and informs the owners of each unit of their schedule. Additionally, the ISO sets an hourly market price for both energy and ancillary services. The ISO sets the hourly price based on the selection of the most expensive unit required to meet load obligations as well as transmission considerations. The mechanics of the setting of the market price are discussed in remarks at the close of this section.

Retailers pay the hourly market price to the ISO for the demand that they have indicated. In turn, the ISO pays the market price to suppliers. Not that the demand is reduced by the interruptible supply that the retailers indicate is available. Also not that all selected energy producers receive the same payment for the energy that they provide, even if their bid is lower than the market price.

Additionally, the ISO selects ancillary service providers from among the ancillary service bidders. A market price for the capacity component of the services is determined, and every selected bidder receives the market price. If called upon, in real time, the provider receives an energy payment as described in Activity 7.

Activity 6: Intraday Markets

Intraday markets allow both retailers and producers to adjust their day-ahead schedules at predetermined times during delivery day. Both retailers and producers are interested in making adjustments as they update their delivery day requirements and output capabilities with information available throughout the delivery day. Retailers can sell back supply to the ISO or purchase additional supply as actual demand deviates from the day-ahead forecast. Similarly, producers may repurchase supply from the ISO or sell additional supply into the market. A market price is set by the ISO, which optimizes balancing resources that producers and retailers bid into the intraday markets.

The following is an illustrative example. In this example there are four market players, two retailers and two producers. For a particular hour, Retailer A has scheduled 700 MWH of power day ahead and Retailer B has scheduled 500 MWH. The market clearing price in the day-ahead market is $70. In the intraday markets Retailers A and B adjust their load forecasts. For a particular hour, Retailer A determines that it requires an additional 20 MWH while Retailer B determines that it has a surplus of 10 MWH. Retailer B bids in a sale of 10 MWH to the ISO at a price of $75. Producers C and D have been scheduled to deliver 400 MWH and 300 MWH, respectively, in the day-ahead markets. Intraday, Producer C bids an additional 10 MW into the system at a cost of $80/MWH while Producer D submits a bid also with a capacity of 10 MW at $85/MWH. The least-cost rescheduling choice for filling the 20 MWH position of retailer A is to purchase 10 MWH from retailer B and 10 MWH from producer C. The ISO sets the market clearing price at the marginal cost of these bids, $80. Retailer A pays $80/MWH for the additional power, while players B and C receive $80/MWH.

Activity 7: Real-Time Balancing and Imbalance Penalties

Real-time balancing is achieved through the regulation and reserve services that are bid into the ISO in the day-ahead markets. The ISO selects units that are placed on AGC as well as units to provide primary and secondary reserves in the day-ahead market, as described in Activity 4. During delivery day, the requirement may be increased and additional units scheduled.

Large deviations in load or forced outages require the dispatch of primary and secondary reserves. The ISO calls upon these services in real time as needed. The ISO dispatches ancillary service by selecting the set of resources that can provide required services at the least costly energy price as bid into the ancillary service markets. The most costly of all the ancillary service prices determines an energy payment that all ancillary service providers receive.

Both retailers and suppliers pay penalties for deviating from their schedules. Penalties are based on ancillary service prices. Rewards for ancillary services are paid from the penalties that are collected to the ancillary service providers. The penalty is referred to as an imbalance penalty.

Most short-term markets operate very similarly to the hypothetical market, although there are differences. PJM does not run an intraday market. Instead, there is a real-time price that is applied to the difference between a market participant's day-ahead schedule and real-time position.

For example, suppose that a retailer scheduled 150 MWH of electric energy for the hour ending 6 A.M. in the day-ahead market with a price of $52/MWH. Suppose that the retailer's actual consumption is 125 MWH and the real-time price is $46/ MWH. The difference between the scheduled demand and actual take is 25 MWH, which acts as a supply against the original 150-MWH demand. The settlement between the ISO and the retailer is the following. The retailer pays the ISO $52/ MWH for 150 MWH of electric energy. Then the ISO pays the retailer $48/MWH for 25 MWH of power that the retailer did not consume. The net result on the 25 MWH of overscheduled demand is that the retailer pays $100 for the over-scheduling, $4 for each of the 25 MWH.

In PJM the real-time price is calculated by the energy-only cost of ancillary services that are called upon to make up differences between day-ahead schedules and real-time deliveries and demand. Typically, although not always, when the day-ahead schedule undersupplies the real-time market, the real-time price is above the day-ahead price. Conversely, when the day-ahead schedule oversupplies the real-time market, the real-time price is below the day-ahead price.

We close this section with many remarks.

Remarks

- The economic dispatch that the ISO determines is optimal with respect to the producers' bids. It is optimal with respect to the operating conditions of the supply stack only if producers bid in their actual operating characteristics. Under competitive conditions, this seems to hold. However, producers may bid well above their operating characteristics. Typically, there are regulatory mechanisms to ensure that producers do not bid excessively beyond their operating costs. Such mechanisms include price caps and investigations into pricing behavior with penalties for exercising market power. There has been discussion of allowing the ISO to contract for generation and act as a supplier of last resort. The ISO would be required to bid its operating characteristics that act as a cap on prices.

- Note that all producers receive the market clearing price for their generation. All units selected for dispatch have bid costs at or below the market price; most bids are below. This provides a profit margin equal to the difference between the actual variable cost of production and the market clearing price. An issue is whether or not this revenue is sufficient to cover fixed costs and provide a profit to producers under competitive conditions so that producers have incentive to construct new units. This issue is further addressed in the following sections on midterm market structure and long-term market structure.
- It is critical that the standard for bidding into the energy markets includes adequate representation of the operating parameters of the units. Otherwise, producers will have to pad their bids in order to be able to ensure cost recovery of their variable costs. For example, suppose that ramp-up rates are not included in the bidding process. Then the ISO may dispatch a unit beyond its ramping capabilities. The producer will not be able to meet its schedule and will have to pay an imbalance penalty. To ensure enough revenues to cover the imbalance fee, the producer will increase its price.
- Another example is the start-up cost. If the start-up cost is not included in a bid, producers will have to recover this cost through a $/MWH charge. But before the dispatch being set, producers don't know how many MWH a unit will be called upon to produce. The producer will not have the information required to determine the additional charge required to cover the cost. The producer will pad its bid to ensure that this cost is recovered.
- There are more similar examples. These illustrate the point that unspecified operational information leads to more uncertainties and potentially higher prices.
- An issue that we have not addressed is the management of transmission information. The ISO is the only agent authorized to have complete access to transmission information. If any producer is able to determine the transmission constraints, day ahead or in real time, the producer could then use the information to adjust bids for its supply. For example, suppose that a producer knows that a transmission line used to import power to a substation where its generator is connected to the grid has tripped. The producer then recognizes that it has market power and can increase its price offering.

  The ISO may not make exclusive transmission information available to any single market participant. All information that it releases must be made through a public source.

- In addition to an energy imbalance penalty, markets also impose a capacity imbalance penalty. For purposes of reliability, markets often have a regulatory requirement that retailers demonstrate the purchase of sufficient capacity commitment to meet the load requirements of their customer base. Failure to demonstrate this commitment results in a capacity penalty. Capacity markets are discussed further in the next section.
- The penalties are a contentious issue. On one side of the issue are those who believe that penalties cause higher prices to end customers as retailers pass these costs along. This side believes that real-time settlement as with PJM addresses balancing. On the other side are those who argue that penalties ensure that the best information set for setting an economic dispatch is available to the ISO. Additionally, the penalties are necessary for the ISO to be effective in its role as steward of reliability. Those in favor of penalties point to the experience in California. Indeed, without sufficient penalties energy companies misinformed California's ISO about their load forecasts. This is further discussed in Chapter 11.
- One feature that has not been addressed is the mechanism for providing reactive power. Indeed, all discussions to this point have only addressed the dispatch of real power. Ensuring the delivery of reactive power is the responsibility of the ISO. The ISO must ensure that the dispatch it sets includes sufficient local resources to provide reactive power. Additionally, the ISO activates capacitor banks and other reactive power generators as required to maintain grid integrity. The costs for this service are paid by retailers in the form of grid service payments.
- A common method for determining the market clearing price is known as locational marginal pricing. In this setting there is a price determined at each bus of the network. For supply buses, those at which units interconnect into the grid, the market clearing price is that of the dispatched unit with the most expensive bid. At all other buses, the cost is that of the marginal MWH of load. At a single load bus, one calculates this as follows.
    - Increment the load at the bus one wishes to price by a single MWH.
    - Determine a new dispatch for the required additional MWH.
    - Calculate the new dispatch cost across the network: the summed bids of all dispatched units.
    - Subtract the cost of the original dispatch from that of the new dispatch to determine the load-sensitive component of the market clearing price.
    - Decrement the total transmission transfer capability of all lines in which there is incoming flow by a single MW. Distribute the single-MW decrement to each line as a percentage of the incoming flow through the line.
    - Determine a new dispatch with the reduced transmission capability.
    - Calculate the new dispatch cost across the entire network: the summed bids of all dispatched units.
    - Subtract the cost of the original dispatch from that of the new dispatch to determine the transmission-sensitive component of the market clearing price.

- Sum the load-sensitive component and the transmission-sensitive component to arrive at a total market clearing price for the bus.

One feature of this calculation is that it requires producers to bid in their units in which prices do not incrementally decrease as MW output increases. For example, when the output is between 150 and 200 MW and the bid is $85/MWH, the bid must be at or above $85/MWH whenever the output is above 200 MW. Indeed, incrementally nondecreasing bids is a bidding rule in markets that adopt locational marginal pricing. This rule is not consistent with actual operational costs since there are output regimes in which the incremental heat rate decreases as output increases. An example of decreasing incremental heat rates was provided in Example 3, Section 3.5.

7.3 LONG-TERM MARKET DESIGN: NO CLEAR SOLUTIONS

The task of developing market-based incentives to address long-term supply-demand equilibrium is more difficult for power than for other goods and commodities. There are two central causes for the difficulty. The first is the localized nature of power production. Power plants require large capital outlays, but they are constructed to service local markets. Local markets do not provide the stability of demand that national and global markets provide. In this environment investors prefer guarantees that markets are not able to provide. (Regulators are also concerned about localization because of the potential that production ownership may be concentrated among a few producers. The concentration of ownership constrains competition.) A second cause of difficulties comes about as a result of reliability issues that surround electricity. Reliability requires excess capacity to provide reserves. It is difficult to establish a pure market-based mechanism that both provides incentives for reserves and establishes appropriate reserve margins. As of yet there is no consensus on the structuring of long-term markets.

Another issue that markets must grapple with is the upgrading of transmission equipment associated with the construction of new power plants. As noted in Chapter 5, such upgrades are costly. To maximize the cost-effectiveness of the upgrade, future grid requirements beyond those required to accommodate the new power plant are considered. Financing, cost allocation, and cost recovery of the upgrades must be addressed.

This section presents a discussion of project planning, followed by the concerns of different parties along with proposals that are under consideration to address concerns.

7.3.1 Project Planning

Project planning in a market environment proceeds in a very similar fashion as it does in a utility environment with some differences. The steps mapped out in Section 5.1 apply; as with a utility environment, there is a host of regulatory agencies that must approve projects and the list of approving agencies is much the same as in a utility environment.

One difference is the requirement of a system impact study to address transmission upgrade requirements. It is the responsibility of the ISO to perform the system impact study as well as the remedial actions that are necessary to maintain grid reliability. A cost recovery mechanism for the upgrades must be agreed on through a regulatory agency, usually the utilities commission of the state in which the proposed project is located.

7.3.2 The Financial Environment

Producers and banks that provide financing prefer long-term contracts that cover their cost of capital and provide an established profit as opposed to relying on short-term and midterm sales to generate profits. Short-term and midterm markets are not considered reliable income sources. While power plant financing extends over 20- to 30-year time frames, midterm markets have only a limited degree of liquidity over a 1-year time frame. Day-ahead markets are very uncertain because of localization. Overconstruction within a location is of concern since there are limited exporting opportunities. Overconstruction can cause a price collapse; both output and profit are reduced. A project cannot rely on exports because transmission limitations as well as the additional cost of transmitting to a neighboring region affect both production volumes and profits.

Banks and producers desire appropriate counterparties in order to secure long-term contracts. Unfortunately, retailers prefer not to enter into such long-term contracts. Their main concern is that they do not have a stable customer base. It is very difficult for retailers to estimate customer requirements over a long term even if the customer base remains fixed. In a competitive market where the customer base may change, retailers are far less sure about their long-term needs and are unwilling to make long-term commitments. Structuring long-term contracts often requires the intervention of a state regulating agency that provides guarantees of cost recovery through a local utility or the ISO. Of course, this approach contradicts the goal of developing free markets.

There are no simple solutions to these problems. Proposals to motivate new construction include the following.

- Exit fees: To maintain the customer stability of retailers, require that customers switching retailers pay an exit fee to the company that they originally signed up with. Although this provides customer stability, it contradicts the whole purpose of moving away from traditionally regulated markets, providing customers with choice. Under this scheme, customers must pay to have a choice. Additionally, it is uncertain that this solution would create stability of the customer base over the time horizons required for financing generation, 20 years.
- ISO ownership: Another solution is to allow the ISO to purchase and own capacity as the steward of reliability. The ISO could recover the cost of

ownership through its standard ISO charges. This increases the market role of the ISO considerably. It also increases regulatory legislation and oversight. Additionally, this increases the percentage of overall end costs that are not subject to competition.

- Producer ownership with guarantees from the ISO: This solution allows producers to own and operate plants. The ISO makes contractual commitments to provide payments that make up the difference between revenues from the short-term market and capital cost requirements. In return, the producers provide guaranteed bidding parameters into the short-term markets over the lifetime of the plant. This solution is a variation of the previous one and has similar effects.
- Producer ownership with guarantees from local utilities: Under this solution producers own and operate plants, but utilities lease them. The utility pays all fuel costs, bids the output into the ISO short-term markets, and receives all revenues from the ISO. Additionally, the capacity certificates belong to the utility. Utilities only agree to these conditions if they receive regulatory protection in the form of rate guarantees and limitations on customer choice. This solution conflicts with the goal of establishing free markets.
- Market price signals: Develop a market with energy-only prices and allow the price signal to determine new construction. The proposal is to provide a price signal that both consumers and producers respond to. Producers respond to the price signal by building new generation. Drawbacks with this include the following:
    - Because of the time required to build new generation, time delays in the response to high price signals are significant. In the interim, because of transmission limitations a location cannot tap national or international supply sources, as are available in other markets.
    - The system provides no mechanism for setting reserves and guaranteeing reserve margins. Reliability may be impaired.
    - Banks and producers don't support this model. For the reasons given above, they will not promote construction without long-term contracts.

7.3.3 Transmission Upgrades

System upgrades to the transmission line are very costly. In the event that an upgrade is necessary to accommodate a new power plant, substantial oversizing of the upgrade may occur to attend to future needs. Restating an example from Chapter 5, assume that studies indicate that a proposed power plant will cause power flows across a transmission line to exceed the line's limit by 10 MW. It would be foolish to go through the expensive process of resizing the line by only 10 MW. Instead, resizing the line by a magnitude of several hundreds of MW to meet future needs and increase reliability is common.

In a utility environment, the additional costs of the transmission upgrade are allocated to the project and are later recovered through regulated customer rates. There is far more complexity in a market environment. The reality of interdependencies between grid equipment and generation equipment cannot be dismissed, and three parties must agree on a plan to proceed: the IPP wishing to construct the project, the owner of the transmission equipment affected by the upgrade, and the ISO. The issue became an impediment to construction of new generation. In response the Federal Energy Regulatory Commission (FERC) issued a regulatory process for allocating upgrade costs and providing cost recovery. The process relies on the ISO to determine upgrade requirements. It is then the responsibility of the IPP to perform the upgrades before its newly constructed unit comes on line.

The FERC process does not specify a methodology for determining upgrades and associated costs, and the outcome is not at all a good indicator of actual transmission costs associated with a project. An ISO may use its own judgment and require an IPP to finance grid expansion beyond the requirement of the IPP's specific project. The FERC regulation is an administrative process that provides a pathway for project completion.

7.4 MIDTERM MARKET DESIGN

We next move on to midterm market design. Both retailers and producers have more difficult issues to address in midterm markets than utility planners must consider in a traditional utility environment. They must use midterm markets to financially balance their forecasted positions. Retailers must purchase supply to match load commitment, and producers must sell energy and capacity. By contrast, midterm planners in traditional utilities are working with well-defined portfolios that are not far out of balance. Their main task is optimizing their portfolios to meet maintenance requirements and energy constraints as opposed to actual portfolio creation.

In this section we present the market instruments that both retailers and producers use to balance their portfolio positions. Afterward, we discuss the market's ability to address midterm operational issues, maintenance scheduling, and use of energy-limited resources.

7.4.1 Market Instruments

There are two components of supply that markets must address: energy and capacity. Energy is the energy content of electricity that is delivered to a specified bus. Capacity is output availability of physical units. In this subsection we concentrate on wholesale financial products that producers and retailers utilize to manage financial risk. Aside from producers and retailers, financial institutions are also market participants in midterm markets. Financial products are contracts that do not result in the physical delivery of energy or the physical exchange of capacity between the counterparties. They are contracts that fix the cost of supply for both products through financial settlement. This section provides a description of common financial products.

Energy Markets

We begin with an example to demonstrate how financial products fix the costs of physical transactions without an exchange of the physical products. Common financial instruments are then described.

EXAMPLE *Financial Hedging with Fixed for Floating Contract*

Suppose that a retailer forecasts a need for 100 MW of baseload energy during a given month and the retailer wishes to fix the price of its supply instead of risking an uncertain price in the short-term markets. The retailer can contract a financial settlement with a third party that works as follows. There is a fixed contract price. Whenever the short-term day-ahead market clears below the fixed contract price, the retailer owes the difference to the third party. Whenever the short-term day-ahead market clears above the fixed contract price, the third party owes the difference to the retailer. In this way the retailer effectively pays the fixed contract price for the amount of the contract, provided that it actually schedules the contract quantity with the ISO.

Let's put some numbers to this example. In January of a given year a retailer and a third party contract for 100 MW of baseload to be settled against the ISO's day-ahead index at a specified delivery point with delivery time July of that same year. The fixed contract price is \$83/MWH. During the month of July the actual average price of baseload is \$79.24/MWH. We refer to this as the monthly index price. The retailer must pay the third party \$3.76/MWH for every MWH of the contract. Not that on scheduling this energy for delivery with the ISO, the average price that the retailer pays to the ISO is the day-ahead market clearing price of \$79.24/MWH. Accordingly, the retailer has set its total procurement cost at \$83/MWH through the contract with the third party.

If on the other hand the average July market clearing price for baseload is \$85.36, the third party owes the retailer \$2.36/MWH for every MWH of the contract. On scheduling this energy for delivery to the retailer, the average price that the retailer pays to the ISO is the day-ahead market clearing price of \$85.36/MWH.

Figure 7.1 illustrates the payment between the retailer and counterparty as a function of the ISO index that settles in July. A positive payment indicates that the seller pays the retailer, while a negative payment indicates that the retailer pays the seller. This example shows that, in either case, the retailer's procurement costs are set at \$83/MWH through the contract with the third party irrespective of the outcome of the ISO index.

In the above example the third party could be a producer, another retailer, or a financial institute that holds no physical positions in the market. Suppose that the third party is a producer and that the producer forecasts at least 100 MW of baseload sales into the market. Then the producer sets its sales price with the contract.

A list of financial energy products follows. The list is similar to what is available and quoted by brokerage firms. Typically the liquidity for these instruments does not go beyond 1 year, with much less liquidity for options. The instruments trade in

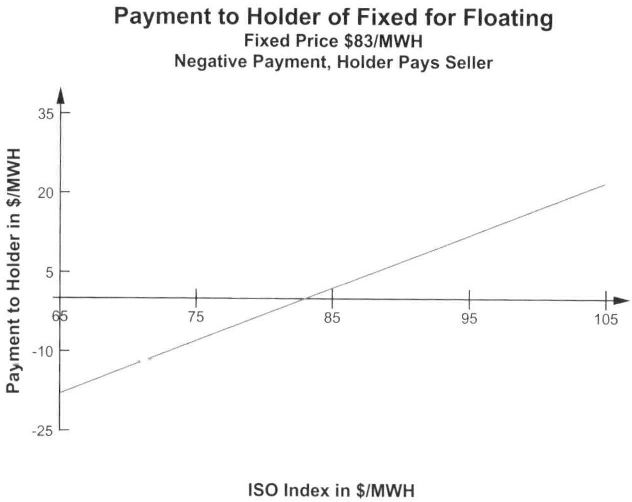

Figure 7.1

monthly blocks, or strips of monthly blocks and counterparties enter into contracts over monthly time frames. In the above example, the contract was signed in January and settlement of the contract depended on prices over the entire month of July. Once into delivery month contracts trade for the balance of the month. Contracts trade in block sizes that vary by market. Typically the block size is 25 or 50 MW. Also, the contracts trade for peak, off-peak or baseload time frames. In some markets superpeak hours are traded.

The instruments presented below all have the feature that they settle against the ISO's day-ahead market clearing price. Day-ahead markets are preferred to hourly clearing markets because without physical power financial players do not participate in hourly markets. The ISO makes its day-ahead price publicly available, allowing for financial settlement. A list of instruments follows.

1. Fixed for floating contract. This is the product that is described in the above example. The fixed-price component of the fixed for floating contract is viewed as the value of electric energy for the contract time frame.
2. Monthly indexed call option. Associated with all options is a strike price. The seller of the monthly indexed call option must pay the purchaser the difference between the ISO monthly index and the strike price whenever the index settles above the strike price. If the ISO monthly index settles below the strike price, there is no transfer of payment between the seller and holder.
3. Daily indexed call option. For every day during settlement month that the ISO index is above the strike price, the seller of the option must pay the purchaser the difference between the index and the strike price. For every day that the ISO index settles below the strike price, there is no transfer

payment between the option's seller and holder. Note the difference between the daily and monthly options. Settlement duration is the same for each. However, the daily option is settled each day over settlement month, whereas the monthly option is settled only at the end of the month over the aggregate monthly ISO index. Note that for the same strike price, the payment from seller to holder of a daily option is always greater than or equal to the payment from seller to holder of a monthly option. It is easy to construct examples in which the holder of the daily call option receives payment but the holder of a monthly option does not receive payment.

4. Monthly indexed put option. The seller of the monthly indexed put option must pay the holder the difference between the ISO monthly index and the strike price whenever the index settles below the strike price. If the ISO monthly index settles above the strike price, there is no transfer of payment between the seller and holder.
5. Daily indexed put option. For every day during settlement month that the ISO index is below the strike price, the seller of the option must pay the purchaser the difference between the index and the strike price. For every day that the ISO index settles below the strike price, there is no transfer payment between the seller an buyer. Note the difference between the daily and monthly options. Settlement duration is the same for each. However, the daily option is settled each day over settlement month, whereas the monthly option is settled only at the end of the month over the aggregate monthly ISO index. Note that for the same strike price, the payment from seller to holder of a daily option is always greater than or equal to the payment from seller to holder of a monthly option. It is easy to construct examples in which the holder of the daily call option receives payment but the holder of a monthly option does not receive payment.

Figures 7.2 and 7.3 provide graphs of the payment transfer from an option's seller to an option's holder. Within each graph a strike price is set, and the graph represents the payment transfer as a function of the realized ISO index. The horizontal axis gives the ISO index, and the vertical axis provides the payment. In addition to the payment transfer provided by the graphs, there is also a set payment that the buyer of the option pays the seller of the option. The set payment is paid on establishing the contract and is called the contract premium.

Markets also trade physical instruments through brokers. The difference between physical and financial instruments is that instead of a financial settlement between counterparties there is physical delivery of electric energy. The physical counterpart to the fixed for floating contract is a forward contract in which the purchaser pays the seller a set price for a specified energy quantity and the seller delivers the energy quantity to a specified location throughout the delivery month. Physical daily calls are similar to financial calls except that the seller delivers electrical energy whenever the holder requests delivery and the holder pays the strike price as opposed to the day-ahead market price. Physical monthly calls differ from financial monthly calls in that the holder of the option decides whether or not to call on the seller for delivery

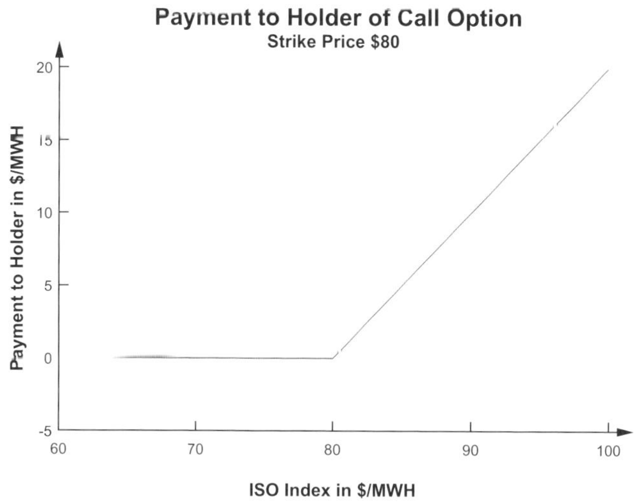

Figure 7.2

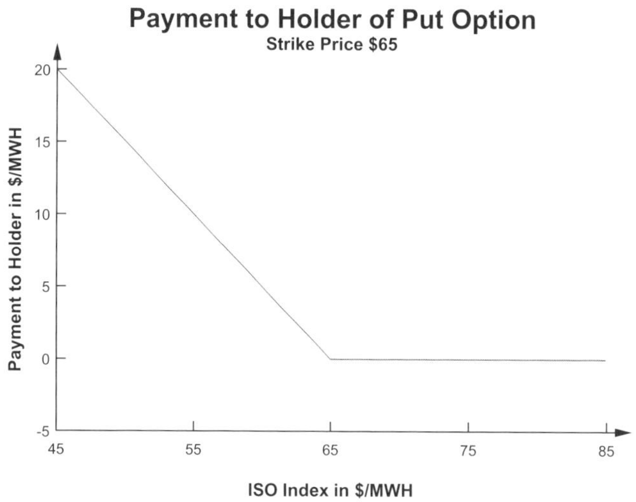

Figure 7.3

of electrical energy on the last trading day of the month before delivery date. With physical puts, the seller of the put is obligated to purchase electricity from the seller at the strike price if the holder of the put chooses. The holder of the put makes the decision on the day before delivery date in the case of daily puts and on the last trading day of the month before delivery date in the case of monthly options.

Except for monthly options, the financial implications of physical and financial instruments are similar. As an example, consider the daily call. The holder of the option requests physical delivery whenever the broker-traded market price is above the strike price; otherwise, the holder makes a purchase for a cheaper price in the market. Whenever the holder requests delivery, the value to the holder is the difference between the market price and the strike price.

Brokers and trading companies initially hoped to develop liquid markets for a wider array of contracts. Such contracts include options in which the settlement depends on prices of both power and fuels, instruments in which the settlement depends on prices of other commodities such as aluminum, and even instruments in which the settlement depends on weather outcomes. These sorts of products have never become commonly traded.

Counterparties also sign contracts outside of broker markets through direct negotiations. The contracts of greatest interest are options in which settlement depends on prices of power and fuels. Contracts offering load-following services to account for daily as well as hourly load fluctuations are also of interest. Although these contracts do not trade in the broker markets, their prices depend on the prices that are observed in the broker markets.

Capacity Markets

In some markets, aside from financial instruments for energy there is also a financial instrument for capacity. The capacity market comes about through a regulatory statute that requires retailers to demonstrate sufficient capacity coverage for their customer base. Coverage is demonstrated in the form of a certificate. Certificates for capacity are initially issued to producers that can demonstrate ownership of generation and in some circumstances ownership of generation coupled with rights to transmission. The certificates are provided in monthly time frames, similar to the financial instruments.

The certificates may be sold to other parties. Transfer of a certificate does not mean the transfer of ownership of the underlying power plant. However, as mentioned above, retailers must have sufficient certificates to demonstrate capacity coverage for their customer base. Recall from the previous section that retailers incur penalties for not meeting their capacity coverage. Eventually, markets transfer the certificates from producer to retailer, although they may go through financial intermediaries. Just as with energy markets, liquidity for capacity markets does not extend beyond 1 year.

There are two related purposes for the capacity requirement. The first is to address reliability. The requirement ensures that, in aggregate, retailers do not make commitments to deliver more power than the available capacity. The second purpose is both a financial and a reliability issue. Market designers hope to keep indexed prices close to the true system lambda of an economic dispatch. However, the revenues a producer earns from these prices are not sufficient to cover fixed costs and earn a return on a power plant. The capacity payment is meant as a mechanism from which producers can earn a revenue stream to ensure financial viability of their plants. Without financial viability plants would close; this would impact reliability.

7.4.2 Maintenance Scheduling and Energy-Limited Resources

Recall that two operational issues of midterm planners in a utility environment are maintenance scheduling and planning the use of energy-limited resources. In this subsection we examine the ability of the markets to address these issues. We begin with maintenance scheduling.

Recall that there is limited flexibility in preparing a maintenance schedule; personnel and materials must be prepared well ahead of planned maintenance. The problem of optimal maintenance scheduling was posed in Chapter 6; this is a least cost of service problem. In a market environment the objective is to maximize the net present value (NPV), present value of future earnings, of the portfolio subject to meeting scheduling requirements.

It is very difficult for the market to coordinate maintenance scheduling across many different producers. Since maximizing NPV is an objective, all producers want to schedule maintenance during the same month, the month with the least value. To be able to perform the task of coordinating outages a market must be very liquid over at least 1 year's time frames since maintenance schedules are typically set 1 year in advance. Instruments in these very liquid markets would have to respond to maintenance schedules by increasing value during months in which maintenance is overly prescribed, providing an incentive to shift maintenance to other months. Currently, markets do not have sufficient liquidity to provide appropriate market signals a year in advance. Typically, markets do not sell monthly fixed for floating instruments a year in advance of delivery. At best, a strip of 4 months is available.

An additional requirement for coordinating maintenance schedules is that the market must provide intramonth information so that producers do not overschedule maintenance during a particular week. Unfortunately, the market provides no intramonth information.

The problem associated with coordinating maintenance scheduling has stirred a debate that is similar to that on meeting reliability requirements. On one side of the debate are those who look to the ISO to coordinate maintenance schedules. This would provide the ISO with additional regulatory oversight responsibilities. On the other side of the debate are those who prefer to let the markets organize to take care of the problem.

Although the markets have not addressed maintenance planning sufficiently, there has been one success. Ensuring availability of capacity during peak demand months has taken on a greater urgency than existed in a pure utility environment. Accordingly, producers rarely schedule maintenance during peak demand months. In line with this success, producers also make every effort to ensure that energy limited resources are available during peak demand months. The market provides economic incentive to ensure availability.

Chapter 8

Asset Management in Short-Term Markets

In this chapter we present the market interactions of retailers, producers, and integrated energy companies in the short-term markets. The market participants bring their physical positions into balance by providing load and supply bids, which the ISO uses to set a balanced schedule. The market participants bid with the objective of maximizing the profit of their portfolios. This chapter provides a description of the bidding decision for each of the market participants. Sections 8.1 and 8.2 present retailers' and IPPs' bidding decisions for a portfolio without market instruments. Section 8.3 presents an integrated energy company's bidding behavior and relates this bidding behavior to that of a producer with a portfolio containing market sales.

8.1 RETAILERS

As noted above, the physical position of the retailer is brought into balance in the short-term markets. The process begins in the day-ahead market and continues through to delivery time. We first present a description of the retailer's position. Afterwards, retailer behavior for a simple portfolio without market instruments is examined.

8.1.1 Description of a Retailer's Portfolio

A retailer's portfolio consists of a set of customer contracts that obligate the retailer to provide power along with network payments, imbalance penalties, and market instruments. The retailer is interested in the cash flow, which is the sum of cash flows from each component of the portfolio. The following equation provides a representation for the cash flow over a single delivery hour, specifying each component:

Electric Power Planning for Regulated and Deregulated Markets. By Arthur Mazer

$$R_h = L_h(p - n) - \underline{L_h}P_h \pm D_h - I_h + a_h(P_h - f)$$

- R_h represents the retailer's cash flow for delivery hour h.
- L_h represents the actual load in MWH of the customer base for delivery hour h.
- p represents the average contract price with the retailer's customer base in \$/MWH.
- n represents the network cost in \$/MWH.
- $\underline{L_h}$ represents the load that the retailer requests from the ISO for delivery hour h in MWH.
- P_h represents the ISO day-ahead market clearing price in \$/MWH for delivery hour h.
- D_h represents the profit or losses from participation in the delivery day markets for delivery hour h.
- I_h represents imbalance penalties.
- a_h represents the volume in MWH of fixed for floating products that the retailer has purchased for delivery hour h.
- f represents the average fixed cost in \$/MWH over all fixed for floating contracts in the retailer's portfolio.

Variables represented by capitalized letters are unknown before the clearing of the day-ahead market, whereas variables represented by small letters are known quantities. Terms that are positive indicate incoming revenues, whereas terms that are negative indicate costs that the retailer must pay. The retailer earns a profit by ensuring that the term L_hp is high enough to cover all the portfolio costs.

Note that the delivery date revenues represented by D_h may contribute to profit or may be a cost. This term is further addressed in Section 8.2.

8.1.2 A Portfolio Without Market Instruments

Without market instruments, the retailer's cash flow is represented as follows:

$$R_h = L_h(p - n) - \underline{L_h}P_h \pm D_h - I_h$$

The retailer's objective is to maximize the profits at delivery time. The only positive contribution to the retailer's profit comes from the fees that the retailer collects from its customer base. All the other components are costs. The retailer must optimally bid in load requirements to minimize these costs.

There are two components to the load bid, the uninterruptible component and the interruptible component. Because the retailer is committed to providing power to its uninterruptible customers regardless of the ISO index price, the retailer wishes to minimize imbalance charges with the ISO and ensure that this load is scheduled. Accordingly, the retailer's bid for the forecast of the uninterruptible load component indicates a willingness to purchase power at any price. Accurate forecasts allow retailers to avoid imbalance charges.

The second component of the load bid is that associated with the interruptible contracts. A load bid coupled with an interruptible contract depends on the contract's terms of interruption. We discuss two types of interruptible contracts: price-driven interruptibility and value-driven interruptibility.

Price-driven interruptible contracts are the simplest type of interruptible contracts. They provide the retailer with the right to curtail power deliveries whenever the ISO index settles above an agreed-upon threshold. The load associated with this contract is bid in at the forecasted load up to the interruptible threshold. Above the interruptible threshold, no volume is bid.

Value-driven interruptible contracts are another type of interruptible contract. These contracts provide the retailer with the right to interrupt power for no more than an agreed-upon number of hours per year. There is value in the right to interrupt a future hour of power deliveries. The retailer must be able to place a monetary value on this right and ensure that the forecasted load associated with the contract is bid in at its full level for prices below the interruptible value. Concurrently, the forecasted load is not bid in for prices above the interruptible value. Table 8.1 summarizes the bidding.

The assumptions for the table are the following:

- Forecasted uninterruptible load: 1550 MWH
- Forecasted load of price-driven interruptible contracts: 45 MWH
Interruption price: $250/MWH
- Forecasted load of value-driven interruptible contracts: 95 MWH
Value of interruption: $325

Forecasting of loads is central to minimizing imbalance charges. Unlike utilities in a traditional environment, retailers do not have a good information set for making quality forecasts. A complete information set would contain the most recent information on load so that load forecasting models could be appropriately updated. Although the ISO provides real-time information on overall load levels at the different buses, retailers do not have real-time information on the consumption of their customer base. The aggregate numbers provided by the ISO include different load types: industrial, commercial, and residential. Retailers must be able to parse these numbers

Table 8.1 Retailer's Bid Table

Hour ending 11 A.M.	
Price in $/MWH	Demand in MWH
<250	1690
250 to <325	1645
≥325	1550

to determine the consumption of their customer base. To make matters worse, the customer base is not stable because customers may change their service provider.

To overcome the information deficit retailers sometimes install meters with real-time monitoring and relaying capabilities on a cross sample of their customer base. It is not possible to install meters on all customers because this is expensive. Retailers have developed models that use the sample metered information along with the ISO information to forecast load use.

Once the day-ahead schedule has been set, retailers continue to adjust their schedules with the ISO as forecasts are updated. Within the delivery day market retailers sell back supply to the ISO if load forecasts decrease, or purchase more supply from the ISO if load forecasts increase. This activity is further described in Section 8.2.

Remarks

- Valuing the right to interrupt a customer is not an easy task. There are models that perform this. The models depend on forecasts of prices as well as probability distributions of prices. These forecasts can be very subjective.
- Note that the description of the portfolio is incomplete. The cash flow also contains capacity positions. Capacity positions do not influence bidding behavior in short-term markets and are considered in Chapter 10.

8.2 POWER PRODUCERS

This section follows structure of the previous section. The portfolio of a producer is presented. We next examine the behavior of a producer in which the portfolio contains no market instruments.

8.2.1 The Producer Portfolio (gas-fired generation)

A producer owns generation. Each generator is an asset in the portfolio. For a single hour, the cash flow for the portfolio is the following:

$$W_h = \sum_{i=1}^{n} W_{hi} + q(G - g_f) + a_h(f - P_h) - \mathrm{LD}_h$$

$$W_{hi} = M_{hi}(P_h - H_{hi}(G + g_i) - V_{hi}) \pm D_{hi} - I_{hi} + A_{hi}$$

- W_h is the cash flow in dollars of the producer's portfolio for delivery hour h.
- W_{hi} represents the cash flow from unit i of the producer's portfolio. A producer may have many different generation technologies. For simplicity, we consider only gas-fired technologies.
- q represents the quantity of fixed for floating gas contracts associated with delivery hour h within the portfolio. The quantity is represented in MMBTUs.

- G represents the daily indexed gas price. As with power, there are gas markets with associated indexed prices.
- g_f represents the fixed cost of a fixed for floating gas contract. There are other market instruments available, but we consider only fixed for floating instruments.
- a_h represents the volume of financial instruments in MWHs that the producer has sold associated with delivery hour h. For simplicity, we consider only fixed for floating financial instruments.
- f represents the fixed price of the financial sales in $/MWH.
- P_h is the ISO index for hour h.
- LD_h represents liquidated damages charges. These are charges that the producer incurs when the producer is unable to meet contractual obligations for physical delivery of power to a counterparty. Under such circumstances the counterparty enters into an alternative arrangement for supply and the provider must reimburse the counterparty for the cost of the alternative supply arrangement. Forced outages are a common reason that the producers are unable to deliver as prescribed by their contracts.
- M_{hi} represents the volume in MWH of the producer's bid from unit i that clears the market during delivery hour h.
- H_{hi} represents the heat rate in MMBTUs/MWH of unit i during delivery hour h.
- g_i represents additional costs above the index price of gas associated with transporting gas to the power plant.
- V_{hi} represents the variable operations and maintenance costs associated with operating plant i at output M_{hi} during delivery hour h. This cost depends on operations during previous hours.
- D_{hi} are the profits or losses incurred in the delivery date markets.
- I_{hi} are the imbalance penalties associated with deviations from the scheduled output.
- A_{hi} are the profit accrued from unit i's participation in the ancillary service markets during delivery hour h.

As with the retail portfolio, variables in capital letters are unknown before the ISO sets the day-ahead schedule. The producer portfolio is the supply counterpart of the retailer portfolio. These portfolios are matched in the ISO markets to equilibrate supply with demand.

8.2.2 A Portfolio Without Market Instruments

Without market instruments the producer's portfolio is given by the following:

$$W_h = \sum_{i=1}^{n} W_{hi}$$

$$W_{hi} = M_{hi}(P_h - H_{hi}(G + g_i) - V_{hi}) \pm D_{hi} - I_{hi} + A_{hi}$$

Producers earn profits by selling into the ISO energy and ancillary service markets at prices that are above their production costs. Producers optimize their profits through effective bidding into the ISO markets and efficiently operating their units so as to minimize production costs. Costs that are minimized through effective operations include operating costs, VOM charges, and imbalance payments. Imbalance payments may be incurred by operating the plant at output levels different from the ISO schedule or a forced outage.

Optimal bidding of units into markets is a controversial subject. Features of the power markets cause differences from other markets. Electricity is not storable. It is also essential in modern economies, and there is no substitute available. This causes a certain level of demand that does not respond well to price signals. These features lead to a market in which the bidding behavior of producers changes as operational conditions of the grid change. In this section we investigate bidding behavior and intraday markets.

Producers offer their generation in the day-ahead ISO market with the objective of maximizing profit. Profit maximization requires a balance between profit margin and volume. Under competitive conditions producers are unable to offer units at high profit margins. Producers bid their units to the ISO at costs that are very close to the cost of production. The profit of the marginal unit is insignificant, but units with variable costs below the marginal unit may earn profits. Competitive conditions occur when several producers have surplus generation available; the surplus is with respect to both generation capacity required to meet load and ancillary service requirements.

When conditions are not competitive, offers may be substantially above production costs. This occurs when there are not sufficient numbers of producers with surplus generation capacity.

Extreme conditions occur when generation capacity from every producer is required to meet uninterruptible load as well as reserve obligations. Under extreme conditions, every producer has monopoly power over a portion of the total demand because withholding of generation by any player would mean that total demand could not be met. A regulatory price cap is required because market conditions may become noncompetitive and even extreme. In extreme conditions providers bid a portion of their capacity at cost and a portion of their capacity at or near the price cap. In PJM the price cap is $1000 per MWH.

Figure 8.1 provides a graph of PJM day-ahead prices for the hour ending 4 P.M. during the month of July 1998. That year was marked by capacity shortages and summertime heat waves. From the graph, different pricing regimes are visible. Competitive conditions exist throughout most of the month. Noncompetitive pricing is evident on July 20, 21, 22, and 29. During two of those days, July 21 and July 22, pricing became extreme. There are identifiable thresholds at which noncompetitive pricing occurs. The noncompetitive pricing days correspond to days when heat waves caused high system loads and required capacity that is normally idle as secondary reserves. Price responses as demand passed through this threshold are very

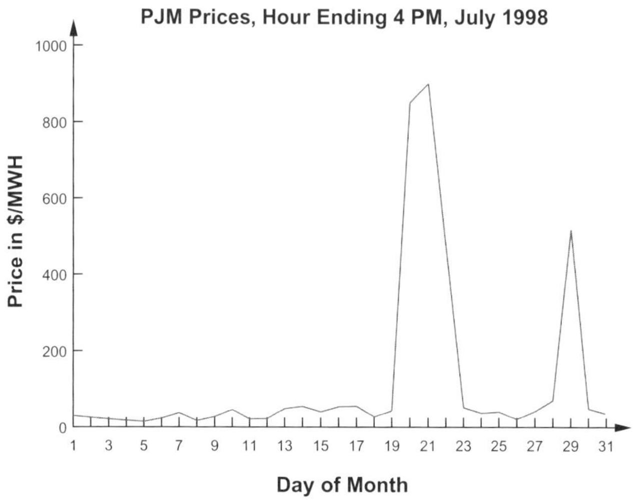

Figure 8.1

dramatic. Equally dramatic is the speed with which prices return to competitive conditions as soon as the supply and load positions become more favorable.

Public data are available for producers to analyze and understand the behavior of market participants and identify noncompetitive pricing thresholds. The data come from several sources. The ISO publishes a history of market clearing prices for energy and ancillary services as well as historical loads. Historical weather is available from various vendors and public sources. Another source of data is the EPA and the National Electricity Reliability Council (NERC). The EPA maintains a data set on emissions of power plants known as the CEMS (continuous emissions monitoring system) data set. The NERC maintains a historical data set on hourly generation output and operating costs by unit.

An effective producer can mine these data sets to understand relationships between weather conditions, load, prices, and units that are dispatched. A producer should simulate the ISO dispatch before entering its bid. This entails performing its own load forecast, setting dispatch prices for all units and performing a simulated dispatch. The simulated dispatch provides the producers with a prediction of the dispatch schedule of their own units as well those of their competition. This also provides a prediction of the market clearing price. Producers may test different bidding strategies against this prediction to determine the most effective strategy. The simulated dispatch informs the producer of the conditions of the market. In particular, the producer is able to determine whether the conditions are competitive, noncompetitive, or extreme and to set offers accordingly.

An issue that producers must address is the bidding of energy-limited resources. For producers, the energy-limited resource is analogous to the retailer's interruptible contract. There is a constraint on the energy-limited resource's overall production. Generating power from an energy-limited resource on any given day restricts the ability to generate from that resource in the future. There is value in preserving the ability to produce energy in the future. The producer must determine this value and ensure that its bid of energy-limited resources produces a profit that exceeds the value.

In addition to bidding into the ISO's energy market, producers also bid into the ISO's ancillary services market. Offers for ancillary service markets include a fee for offering the service and a cost of dispatch in case the service is called upon. As an example, consider an offer for providing secondary reserves. The producer offers its unit at a specific price in dollars. If the unit is selected to provide secondary reserve services, the producer receives the service fee. The producer also provides a start-up cost and output cost with its offer. If the unit is dispatched to produce energy on delivery day, the producer receives the start-up cost and the output cost.

Ancillary service bids associated with energy-limited resources are frequently competitive because this allows the unit to earn revenue with a low probability of being dispatched. To reduce the probability of dispatch, the producer can bid in a high dispatch cost.

Once the ISO sets the day-ahead dispatch schedule, producers focus on two activities: improving their revenues in the intraday markets and ensuring that they do not incur imbalance penalties.

Schedule adjustments arise as load forecasts are updated and retailers foresee less or more demand than was scheduled through the day-ahead market. Additionally, forced outages create a need for producers to alter their day-ahead schedules. Within intraday markets, both retailers and producers can submit bids for altering their schedule. Producers submit bids to decrease or increase their output from the day-ahead schedule. The ISO matches these bids with retailers' updated requirements and sets a new market clearing price.

To determine schedule altering bids, a producer runs scenarios of meeting load obligations in bands that are above and below its day-ahead schedule obligations through an economic dispatch simulator. The economic dispatch simulator finds the cost of altering the bid. A producer uses this cost to determine a bidding price that would improve the producer's earnings when compared to the earnings that the producer would obtain from the day-ahead schedule.

Below is an example that illustrates day-ahead and intraday bidding, selection, and settlements.

EXAMPLE *Delivery Day Adjustments (no transmission losses)*

Suppose that there are three market participants, Retailer A and Producers B and C. Suppose Retailer A has purchased 200 MWH of power from the ISO in the day-ahead market. The hour before delivery, Retailer A wishes to sell back 20 MWH of energy. Suppose that Producers B and C each have sold 100 MWH of electricity in the day-ahead market and each received the ISO index of \$65/MWH. Assume that the imbalance penalty for overscheduling based upon day-ahead AS scheduling is \$15/MWH. With this imbalance fee the retailer's indifference price for the sale is \$50/MWH; in both cases the retailer loses \$15/MWH on the

Table 8.2 Settlements in Day-Ahead and Intraday Markets

		Receives from ISO			Total Payment to ISO
		A	B	C	
Pays to	A		7500	7500	15,000
ISO	B				0
	C	1140			1140
Total amount received from ISO			1140	7500	7500

All amounts are in dollars.

transaction. Assume that the retailer wishes to lose only $10/MWH and bids in a sale of 20 MWH at $55/MWH.

Both producers act as buyers who buy back their production. For both Producers B and C, the indifference purchase price is given by their incremental cost of production. Assume that Producer B's incremental cost of production is $55/MWH while Producer C's is $60/MWH. Furthermore, assume that Producer B bids $53/MWH to purchase back its supply while Producer C bids $57/MWH. The ISO index for delivery hour becomes $57/MWH, and Producer C reschedules with the ISO to deliver 80 MWH of energy instead of 100 MWH.

The cash settlement in this example is as follows. Retailer A pays $65/MWH to the ISO for the day-ahead settlement of 200 MWH. The ISO pays the day-ahead settlement of $65/MWH to producers B and C for 100 MWH apiece. Then producer C pays the ISO $57/MWH for the 20 MWH that is scheduled an hour ahead of delivery time. The ISO makes a payment of $57/MWH to Retailer A for this same quantity.

Table 8.2 illustrates the payment stream in total dollars. The total $15,000 payment that Retailer A pays the ISO is due to the $65/MWH that Retailer A pays in the day-ahead market. The total payment of $1140 that Producer C pays the ISO is due to the 20 MWH that Producer C buys back from the ISO at $57/MWH. The net revenues for any participant are calculated as the difference between the total amount received and the total payment. Note that Producer C earns an additional $3/MWH for the 20 MWH settled in the hour-ahead market above the earnings settled in the day-ahead market, even though Producer C does not deliver that energy. Retailer A ends up paying $8/MWH for energy that Retailer A ultimately does not schedule. However, Retailer A avoids the stiffer imbalance fee of $15/MWH. Also note that the ISO is revenue neutral. All the revenues it receives from market participants are returned to other market participants.

Table 8.3 provides the corresponding volume of energy sales and purchases. For any participant the net volume delivered is the difference between the total volume delivered and the total volume received. With transmission losses the amount delivered and received among all market participants would not net to zero.

Remarks

- Only units that have not been selected to provide ancillary services (AS) may bid into the delivery day energy markets.
- In some markets the ISO also runs a delivery day AS market to adjust AS supply with need. Adjustments may be necessary to meet regulatory commitments in the event of forced outages, as described in Chapter 3.

Table 8.3 Energy Volumes in Day-Ahead and Intraday Markets

		Delivers to ISO			Total amount received from ISO
		A	B	C	
Receives	A		100	100	200
from ISO	B				0
	C	20			20
Total amount delivered to ISO			20	100	100

All amounts are in MWH.

- During the California energy crisis, markets ran amok. Even during off-peak periods when there was a surplus of capacity, producers charged prices significantly in excess of production costs. However, in PJM markets are functioning satisfactorily. Problems with the California market are discussed in Chapter 11.
- The regulated utilities within the Eastern Interconnect utilize the day-ahead wholesale market to optimize their customer costs. Utilities bring their production costs into equilibrium with the market as follows. If a utility's marginal cost to serve its customer load from its own resources is higher than that of the market, the utility will displace output from its more expensive units with a purchase from the wholesale market. Conversely, if a utility's marginal cost to serve its customer load is lower than that of the market, the utility will sell additional production into the market. (The utility's customer base should receive benefits from the sale since the customer base is paying for the fixed cost of the capacity.) Well-managed utilities do not have to rely on wholesale markets because they have their own generation resources that service their customer base; the market is only used to further improve costs. Accordingly, with rare exceptions markets between utilities result in prices that are close to the marginal cost of production.

8.3 INTEGRATED ENERGY COMPANIES

The portfolio of an integrated energy company contains all the elements of both retailers and producers. There are advantages to a combined portfolio that are further addressed in Chapter 10. In this section, we briefly investigate the day-ahead bidding behavior of an integrated company wishing to maximize profits. We then relate this bidding to a producer having a portfolio that contains market instruments.

The integrated energy company has both a demand position that must be filled and a supply position that can produce power. The equation for the portfolio's cash flow during any given hour is the sum of the equations for the retailer's and producer's cash flows. Both the demand and supply positions are cleared through the ISO market. The demand-side bidding of the integrated energy company is identical to that of the retailer. Below is a description of the supply-side bidding.

The supply-side bidding of the integrated company is somewhat different from that of the producer having a portfolio without market instruments. The integrated company must first assess its position and determine whether it will be a net seller or a net purchaser from the ISO. The company is a net seller if the volume of its demand bids that clear the market is lower than its volume of supply bids that clear the market. Conversely, the company is a net purchaser if the volume of its demand bids that clear the market is higher than its volume of supply bids that clear the market. If the company anticipates that it will be a net seller, the company prefers a high ISO index. Alternatively, if the company anticipates that it will be a net purchaser, it prefers a low ISO index.

The company does not want its supply bids to increase the market clearing price in the event that it is a net purchaser. Alternatively, the company does want to maximize profit from energy sales in the event that it is a net seller. One approach toward supply-side bidding that takes into account the position of the integrated company is the following: The company performs an economic dispatch of its own generation resources against a load forecast for its own costumers. The resources that are selected for the dispatch are then bid in at cost. The remaining resources may be bid in order to maximize profit.

8.3.1 A Producer's Portfolio with Market Instruments

If the portfolio of a producer contains sales of financial or physical instruments, the portfolio is similar to that of an integrated company. The bidding behavior is also similar. Referring to the equation that describes a producer's portfolio cash flows, there are two possibilities:

$$\sum_{i=1}^{N} M_{hi} > a_h \quad \text{or} \quad \sum_{i=1}^{N} M_{hi} \leq a_h$$

In the first case, the producer's volume of sales that clears the ISO day-ahead market is greater than the volume of market sales through financial instruments and physical instruments in midterm markets. The cash flow to the producer is similar to that of the integrated energy company when the integrated energy company is a net seller into the day-ahead ISO market. Accordingly, the producer favors high ISO prices and bids just as the integrated energy company does when the integrated energy company is a net seller.

In the second case, the producer's volume of sales that clears the ISO day-ahead market is no greater than the volume of market sales through financial instruments and physical instruments in midterm markets. The cash flow to the producer is similar to that of the integrated energy company when the integrated energy company is a net purchaser from the day-ahead ISO market. Accordingly, the producer favors low ISO prices and bids just as the integrated energy company does when the integrated energy company is a net purchaser.

We close the chapter with some remarks.

Remarks

- Sales of on-peak market instruments with physical delivery create difficulties for producers. The profile of an on-peak instrument is incompatible with operating parameters of generation plants. Operations are constrained by ramp-up and ramp-down rates. However, the on-peak block demands instantaneous power delivery for the first on-peak hour and instantaneous cessation of deliveries at the end of the last on-peak hour. It is difficult to deliver power in accordance with the on-peak profile and avoid ISO imbalance penalties.
- If a retailer believes that the ISO index will settle at a very high price, the retailer may be tempted to cap the volume of its load bid at a certain price, even for the uninterruptible portion. This is risky because the retailer is subject to imbalance penalties on the next day. Additionally, real-time prices have a likelihood of increasing (unless day-ahead prices reach the regulated price cap), because more load appears in the real-time market than was scheduled in the day-ahead market.
- The dramatic price levels that are achieved during excessive pricing periods demonstrate the value of interruptible load to the retailer. When prices reach excessive levels, a retailer may be forced to purchase from the ISO at prices far above the customer's payment to the retailer. The retailer may suffer significant losses. Interruptible load allows the retailer to avoid this outcome.
- The dramatic price levels that are achieved during excessive pricing periods also demonstrate the danger of a retailer's portfolio being far out of equilibrium. A retailer is exposed to excessive pricing for every MWH that the retailer's portfolio is not balanced. A retailer may bring its position closer to equilibrium with the use of market instruments. Risk management is the balancing of a portfolio to mitigate risk. Chapter 10 further addresses risk management.

Chapter 9

Investment Analysis: Long-Term Planning in a Market Environment

In a market environment, infrastructure additions are a result of investment by independent companies seeking profit from the addition. This chapter investigates the current state of affairs for financing and constructing new generation facilities and transmission upgrades. Section 9.1 contrasts the financing of projects in a market environment with financing in a utility environment. In Section 9.2, we present an approach for valuation of merchant plants that is commonly used in the industry. Merchant plants are those that are owned and operated by an IPP in a market environment without regulated cost recovery. Section 9.3 discusses the structuring and pricing of long-term contracts referred to as power purchasing agreements (PPAs). Attention is given to the relationship between valuations of merchant plants and pricing of PPAs.

9.1 INVESTMENT SETTING IN UTILITY AND MARKET ENVIRONMENTS

In traditional utility environments there are two types of projects, utility construction and independent construction. Utility construction refers to a project in which the power plant is owned and operated by the utility. Before construction, projects must be approved by the state utility commission. The approval process includes a provision for the utility to finance the debt and equity of the project through customer rates.

Independent construction refers to projects in which a new power plant is owned and operated by an independent power producer (IPP). Before construction a power purchasing agreement (PPA) is signed between the producer and the utility; the PPA is subject to the approval of the state utility commission. There are many possible structures for the PPA, but the essence is that the utility provides a guaranteed payment to the producer in exchange for agreed-upon power production and capacity. Along with approval for the PPA, the state utility commission provides a mechanism for cost recovery to the utility through customer rates.

The common feature of both of these mechanisms is cost recovery through regulated rates. Furthermore, the recovery mechanism is set in place over the lifetime of the project. Banks find this feature very attractive and readily provide financing for such projects.

In nontraditional markets, neither retailers nor power producers are able to demonstrate regulated cost recovery to banks. Banks are less willing to finance projects that don't have cost recovery mechanisms than projects that are backed by a cost recovery mechanism. In a nontraditional market, the only agent that can provide regulatory assurances is the ISO. This presents a bit of a quagmire for federal and local authorities wishing to develop free markets. To ensure financing for new construction, regulatory agencies must provide the ISO with the authority to enter PPAs so that cost recovery can be ensured. Absent such authorization, financing for new construction may be difficult to arrange and the state cannot ensure reliability of supply. Of course, providing such authorization restores a regulated environment, with the ISO assuming the role of the traditional utility. There is tension between the desire to develop free markets and the desire to regulate reliability. Organizing long-term planning activities is difficult in this environment.

9.2 PROJECT ANALYSIS FOR A MERCHANT PLANT

IPPs select projects that maximize profit. They predict revenues of proposals by performing a simulated dispatch of the proposed power plant against a forecast of fuel and power prices over the time horizon of the project, 20 to 30 years. The simulation dispatches economically against the price forecast as described in Section 4.4. Prices assume the role of the system lambda. The result of the simulation provides a projection of financial and physical outputs, revenues, fuel costs, fuel requirements, and power production. This information together with projections of other costs (maintenance, personnel, capital costs) provides an earnings projection. IPPs then perform a standard net present value (NPV) assessment to determine the value of a project; that is, revenue streams are discounted to a present value with a corporate discount rate.

The earnings and associated value are dependent on the power and fuel price forecasts that are used within the dispatch simulation. Each fuel has its own forecasting methodology and is not addressed; it is assumed that a forecast of prices is available. A consensus approach toward power price forecasting is given below.

Price forecasts rely on two components: an energy component and a capacity component. The energy component is the marginal cost of production, system lambda, from a simulated dispatch. The marginal cost of production is insufficient to cover capital costs of new construction. An assumption that is adopted by the forecasting community is that IPPs will be able to recover the full capital costs, both debt and equity, required to build sufficient resources for ensuring reliability. The capacity price component is the difference between full capital costs and energy payments received by selling electricity at the energy price. All units receive both an energy payment and a capacity payment.

Summarizing the approach, once price curves are established, a given project may be assessed by performing an economic dispatch of the proposed project using the forecasted energy and fuel prices over the time horizon of the project. This provides a cash flow from energy revenues. On top of the energy revenues, capacity revenues are added to the cash flow. The cash flow is then discounted back to a present value with an appropriate discount rate.

To accomplish the above task one needs to understand the present value of future earnings. Conceptually the promise of a future dollar is worth less to an investor than an actual dollar in the present. This is because the promise of a future dollar is uncertain. We illustrate the present value of future earnings with an example.

EXAMPLE *Present Value*

Suppose that an investor has the opportunity to purchase an asset that provides an expected one-time payout of $100,000 one year in the future. What is the current price that the investor should be willing to pay for the asset? The answer depends on market price for the risk associated with the project. In general, the payout is not certain and may be higher or lower. There is a market premium for accepting the risk, and financial analysts use market data to determine the premium. The premium is expressed through a discount rate as follows.

Suppose that the market discount rate for the payout is 12 percent annually. Then the present value of the cash flow is $89,285.71, obtained as follows:

$$100{,}000/1.12 = 89{,}285.71$$

The present value of the cash flow is the amount that an investor should be willing to pay for the asset.

If the asset has an expected payout of $100,000 one year in the future and another expected payout of $100,000 two years in the future, the present value of the project is $169,005.10, obtained as follows:

$$\frac{100{,}000}{1.12} + \frac{100{,}000}{1.12^2} = 169{,}005.10$$

In general, the present value of a stream of expected revenues over a period of years with a given discount rate is the following:

$$\text{PV} = \sum_{j=1}^{n} \frac{CF_j}{(1+r)^j}$$

where

- PV is the present value
- CF_j is the expected cash flow in year j.
- r is the discount rate

Performing an NPV analysis of new construction requires a large set of input assumptions. The input set is the same as that provided in Section 5.3 of the chapter on long-term planning, Chapter 5. Specifics vary between different forecasting techniques; however, one feature is common. Peaking units that set the marginal cost of production recover almost no capital costs from energy sales. They must recover their capital costs through capacity payments. Typically, the full cost of construction for these units sets the forecasted capacity price.

The process is illustrated with the following examples. We use a 12 percent discount rate throughout the remainder of this chapter.

EXAMPLE *A Simplified Energy Price Forecast*

In this example a 20-year price curve is set using the following assumptions.

- A 20-year gas curve is given.
- The forecast is being set for a region where there are no seasonal influences; accordingly, an annual price is sufficient.
- The forecast provides an on-peak and an off-peak price for each year.
- The marginal unit for the on-peak period is a combustion turbine with a heat rate of 10 and capacity of 150 MW. The on-peak price is determined by the full recovery of variable costs, including the start-up cost for this unit. VOM charges are \$3 per MWH, escalating at 2% annually. Start-up costs are \$10,000 per start-up, escalating at 2% annually.
- The weekend daytime price is set by the fuel cost of a CCGT with a heat rate of 7 along with a \$3.50/MWH VOM charge escalating at 2% per year. The VOM charge does not include start-up costs.
- The nighttime price is set by the price at which a CCGT is indifferent toward operating at minimum output or shutting off and assuming a start-up charge the next day. The CCGT has a 2 CT-on-1 steam turbine configuration with a total output of 450 MW. The minimum output of the unit is 150 MW with one CT off line. The start-up cost of the CCGT is \$25,000. The unit's heat rate while operating at minimum output is 10.5, and the start-up cost of each of the CTs is \$10,000.

With these assumptions the on-peak energy price is given by the following formula:

$$\text{On Peak}_j = 10\text{Gas}_j + 3(1.02)^j + \frac{10{,}000(1.02)^j}{150 \times 16}$$

where

- On Peak_j is the \$/MWH on-peak price for year j.
- Gas_j is the \$/MMBTU price for year j.
- 3 is the VOM cost.
- 10,000 is the start-up cost in dollars that is allocated to the 150-MW output over each 16-hour on-peak time period (weekdays from 7 A.M. to 11 P.M.)

The off-peak price is calculated with the following formulas:

$$\text{DT}_j = 7\text{Gas}_j + 3.50(1.02)^j$$

where

- DT_j is the weekend daytime energy price for year j. The formula follows from the assumption that DT_j is set by the gas costs and VOM excluding start-up costs of a given CCGT.

For nighttime operation we have the following:

$$(150 \times 8)(\mathrm{NT}_j - 3.50(1.02)^j - 10.5\mathrm{Gas}_j) - 10{,}000(1.02)^j = -25{,}000(1.02)^j$$

- The above formula equates the total nighttime operating costs, including a start-up charge for the CT that is off line with the start-up cost for the entire plant. The total energy output is 150 MW over 8 hours.
- NT_j is the nighttime energy price for year j in \$/MWH; 3.50 is the VOM charge in \$/MWH.
- 10.5 is the heat rate of the unit.
- 10,000 is the initial start-up cost in dollars of the CT that is turned off while the CCGT is operating at minimum output.
- 25,000 is the initial start-up cost in dollars for the entire plant.

Using this formula one can solve for NT_j. Once NT_j is determined, the off-peak price is calculated by applying appropriate weights to the daytime and nighttime off-peak hours. The result for the off-peak price is the following:

$$\text{Off Peak}_j = \frac{32}{88}\mathrm{DT}_j + \frac{56}{88}\mathrm{NT}_j$$

- 32 is the number of daytime hours during the weekend, 7 A.M. to 11 P.M. over Saturday and Sunday.
- 56 is the number of nighttime hours in a week.
- 88 is the total number of off-peak hours during a week.

Table 9.1 provides a 23-year outlook of on-peak and off-peak prices for a given gas forecast.

EXAMPLE *Capacity Price*

In this example a 20-year capacity price in dollars per kW-year is calculated. The example uses the following assumptions.

- The energy price is provided by the preceding example.
- The capacity price is the price necessary for the CT of the previous example to recover its full construction cost with a discount rate of 12 percent over a 20-year period.
- The construction cost of a CT is \$650/kW, as provided by Section 2.3.8.
- The capacity price increases at a 2 percent annual rate.

Under these assumptions, energy sales from the CT of the preceding example provide no revenues toward the cost of construction. Accordingly, the cost of construction must be recovered through the discounted capacity payment. This provides the following formula:

$$\begin{aligned} 650 &= \sum_{j=1}^{20} \frac{\mathrm{CP}_j}{1.12^j} \\ &= \sum_{j=1}^{20} \frac{\mathrm{CP}(1.02)^j}{1.12^j} \end{aligned}$$

Table 9.1 Price Forecasts

Year	Gas $/MMBTU	On peak $/MWH	Off peak $/MWH
1	7.52	82.51	64.82
2	7.52	82.66	64.70
3	7.25	80.11	62.09
4	6.93	77.06	59.02
5	6.87	76.61	58.34
6	6.84	76.47	57.94
7	6.63	74.53	55.87
8	6.76	76.02	56.96
9	6.90	77.54	58.08
10	7.04	79.09	59.21
11	7.18	80.68	60.37
12	7.32	82.29	61.55
13	7.47	83.94	62.76
14	7.62	85.61	63.99
15	7.77	87.33	65.24
16	7.92	89.07	66.52
17	8.08	90.85	67.83
18	8.24	92.67	69.16
19	8.41	94.52	70.52
20	8.58	96.42	71.90
21	8.75	98.34	73.31
22	8.92	100.31	74.75
23	9.10	102.32	76.22

where

- 650 is the cost of construction in \$/kW.
- CP_j is the current capacity price for year j.
- CP is the current capacity price, $CP_j = CP(1.02)^j$.

Solving for CP gives the following:

$$CP = \frac{650}{\sum_{j=1}^{20} \frac{1.02^j}{1.12^j}} = 75.33$$

The net present value of an investment is the difference between the discounted cash revenues and the initial investment. In this example, the forecasted capacity price provides a net present value of zero for the CT; the discounted revenues from the capacity earnings exactly offset the cost of construction.

EXAMPLE *Net Present Value (NPV) of a CCGT*

In this example we perform an NPV assessment for a CCGT. The following assumptions are used.

- The energy and capacity prices are taken from the preceding examples.
- AS revenues are not included.
- The characteristics of the CCGT are as given in the above energy price example.
- The CCGT is dispatched at maximum output during all on-peak hours.
- There is 1 week's maintenance outage per year. The maintenance cost is covered by the VOM charge.
- There is a 2-month overhaul outage every 5 years through year 20. The cost of overhaul is covered by the start-up charge.
- The construction time of the project is 4 years.
- The cost of construction is \$850/kW as provided by Section 2.3.8.
- The CCGT is in service for 30 years.
- The CCGT receives no net revenues from energy during the last 10 years of its service; any energy payment is offset by costs. However, the CCGT does receive a capacity payment.

To arrive at the NPV we first calculate the present value of the revenues. The revenues consist of energy benefits and costs that are accrued over the first 20 years of operation as well as capacity benefits that are accrued over 30 years of operation.

$$\text{PV} = \sum_{j=4}^{23} \frac{\text{EnergyPayment}_j - \text{FuelCost}_j - \text{Start-UpCost}_j - \text{VOM}_j}{1.12^j} + \sum_{j=4}^{33} \frac{\text{CapacityPayment}_j}{1.12^j}$$

where

- PV is the present value of the cash flow.
- Cash payments start in the fourth year after construction. Energy payments continue for 20 years after construction. Capacity payments continue for 30 years after construction.
- EnergyPayment_j is the total dollar revenue from on-peak power sales during year j.
- FuelCost_j is the cost of gas purchases during year j.
- Start-UpCost_j is the total dollar amount of the start-up costs accrued during year j.
- VOM_j is the variable operations and maintenance cost, excluding start-up costs, accrued during year j.
- CapacityPayment_j is the annual capacity payment in dollars.

We next calculate each component of the revenue stream. Since the off-peak prices are set at the indifference level of operating or not operating during the off peak, the cash flow is the same whether the unit is run during the off-peak or not. As such, we simplify the calculation by considering operations during the on-peak period only with five start-ups per week.

$$\text{EnergyPayment}_j = 450 \times \text{Power}_j \times \text{Hours}_j$$

$$\text{FuelCost}_j = 450 \times 7\text{Gas}_j \times \text{Hours}_j$$

- 450 is the output of the plant in MW.
- Power_j and Gas_j are the power and gas prices, respectively, for year j. These prices in \$/MWH are taken from the energy price example.

- Hours_j is the number of hours the plant operates in year j. We assume 3900 on-peak operating hours for years without overhaul and 3260 hours every fifth year when there is an overhaul.

$$\text{Start-UpCost}_j = 25{,}000 \times (1.02)^j \times \text{Start-Up}_j$$

- $25{,}000(1.02)^j$ is the cost in \$/start-up for year j.
- Start-Up_j is the number of start-ups in year j. We assume 255 for years without an overhaul and 215 for years with an overhaul.

$$\text{VOM}_j = 450 \times 3.50 \times 1.02^j \times \text{Hours}_j$$

- 450 is the output of the plant in MW.
- $3.50(1.02)^j$ is the VOM charge in \$/MWH for year j.
- Hours_j is described under EnergyPayment_j.

$$\text{CapacityPayment}_j = 450{,}000 \times \text{CP}_j$$

- CP_j is as in the preceding example in dollars/kW-year.
- 450,000 is the capacity of the plant in kW.

These values are placed into a spreadsheet where the present value is calculated. The net present value provides the difference between the revenues and the cost of construction. The cost of construction is taken at $850/kW as presented in Section 2.3.8.

Table 9.2 provides the results of the spreadsheet. The column entitled "Future value" provides the sum of the revenue streams in which the benefits, energy payment, and capacity payment are positive while the costs, start-up cost, and VOM are negative. The present value is arrived at by applying the discount rate of 12 percent to the future value streams.

Remarks

- The approach described in Section 5.6 on long-term forecasting for utilities, can be used to forecast capacity stacks across the grid. This provides a good indication of capacity requirements across networks. A prudent IPP along with its supporting investment bank would perform such studies to determine optimal capacity additions by region and ensure that any selected project is in line with the study.
- The approach outlined above is an industry standard. As noted above, the result is that IPPs favor the same type of construction, CCGTs. The NPV of the CCGT is positive, whereas the NPV of the CT is zero. Optimal capacity requires a mix of peaking units along with baseload and intermediate construction. It is difficult for a completely deregulated market to guide IPPs toward optimal generation selection.
- The approach above justifies any project because the result is that any project receives its capital costs including financing and return. Indeed, the approach described above provided the basis for analytics that justified numerous construction projects throughout the Eastern Interconnect between the years 2000 and 2003. There was a lot more construction than necessary, which has caused an oversupply of capacity in the Eastern Interconnect. The assumed recovery of capacity payments has not occurred.

Table 9.2 NPV of CCGT

Year	Energy benefit	Fuel cost	Start-up cost	VOM	Capacity benefit	Future value	Present value
4	135,235,790	(85,135,050)	(6,900,505)	(6,648,840)	36,692,827	73,244,223	46,548,028
5	134,455,076	(84,397,950)	(7,038,515)	(6,781,816)	37,426,683	73,663,478	41,798,636
6	134,206,308	(84,029,400)	(7,179,285)	(6,917,453)	38,175,217	74,255,387	37,620,090
7	130,804,094	(81,449,550)	(7,322,871)	(7,055,802)	38,938,721	73,914,592	33,435,208
8	133,420,176	(83,078,541)	(7,469,329)	(7,196,918)	39,717,496	75,392,884	30,449,921
9	136,088,579	(84,740,112)	(7,618,715)	(7,340,856)	40,511,845	76,900,742	27,731,178
10	138,810,351	(86,434,914)	(7,771,089)	(7,487,673)	41,322,082	78,438,757	25,255,180
11	141,586,558	(88,163,612)	(7,926,511)	(7,637,427)	42,148,524	80,007,532	23,000,254
12	144,418,289	(89,926,885)	(8,085,041)	(7,790,175)	42,991,494	81,607,682	20,946,659
13	147,306,655	(91,725,422)	(8,246,742)	(7,945,979)	43,851,324	83,239,836	19,076,422
14	150,252,788	(93,559,931)	(8,411,677)	(8,104,898)	44,728,351	84,904,633	17,373,170
15	153,257,844	(95,431,129)	(8,579,911)	(8,266,996)	45,622,918	86,602,725	15,821,994
16	156,323,001	(97,339,752)	(8,751,509)	(8,432,336)	46,535,376	88,334,780	14,409,316
17	159,449,461	(99,286,547)	(8,926,539)	(8,600,983)	47,466,084	90,101,476	13,122,770
18	162,638,450	(101,272,278)	(9,105,070)	(8,773,003)	48,415,405	91,903,505	11,951,094
19	165,891,219	(103,297,723)	(9,287,171)	(8,948,463)	49,383,714	93,741,575	10,884,032
20	169,209,043	(105,363,678)	(9,472,915)	(9,127,432)	50,371,388	95,616,407	9,912,244
21	172,593,224	(107,470,951)	(9,662,373)	(9,309,981)	51,378,816	97,528,735	9,027,222
22	176,045,089	(109,620,371)	(9,855,620)	(9,496,180)	52,406,392	99,479,309	8,221,220
23	179,565,990	(111,812,778)	(10,052,733)	(9,686,104)	53,454,520	101,468,896	7,487,182
24–33			—	—	597,018,319	597,018,319	24,441,514
						Sum	448,513,334
						Construction cost year 0	382,500,000
						NPV	66,013,334

- Without the assumed capacity payment, the NPV of both the CCGT and the CT is negative because the cost of construction would be higher than discounted revenues. This demonstrates that energy revenues earned from prices set at the marginal cost of production provide insufficient returns to justify new construction.
- There are several possible sources for capacity payments. One source is the sale of capacity tags, provided that a market exists for the tags. Another source is the sale of electricity into the ISO market when the ISO index rises above the marginal energy cost. There are those who consider price spikes in the ISO index as a necessary mechanism for providing capacity payments.
- A flaw in the approach toward assessing capacity is that the price forecast as well as the subsequent dispatch of projects assume equilibrium. Supply with reserves and demand are in balance. The modeling community is unable to adopt standards for pricing out of equilibrium, nor is it able to adjust the dispatch for such scenarios. There are many difficulties that make the problem nearly insurmountable. Accordingly, the modeling community does not have an approach to this problem; nor should it be expected to. It is important that business decisions are made with a complete understanding of modeling limitations.

9.2.1 Transmission

As noted in Chapter 7, the FERC has instituted a process for allocating costs of transmission upgrades associated with new construction. The cost of the upgrades is initially assumed by the project owner. The owner then recovers these costs through a fixed rate mechanism. It is the responsibility of the ISO to determine the system impact of the new construction along with upgrades that are required to accommodate new generation.

Social interests frequently guide the ISO toward oversizing upgrades. Recall that the cost of upgrades is considerable and there are economic benefits in accounting for future generation capacity additions when considering upgrade requirements for a specific project, even if such future projects are not currently being considered. This means that the owner of a single project may be asked to subsidize expansions for future growth.

While the owner is reimbursed through a regulatory mechanism, the owner still assumes a financial burden. The owner must incur significant debt to arrange loans for financing the upgrades. The debt sits on the owner's balance books until it is paid off. The owner requires compensation for taking on this debt obligation because in doing so the owner foregoes the opportunity to apply the capital to other potentially rewarding projects.

The above NPV calculation does not include the costs of transmission upgrades associated with a given project. The upfront transmission cost should be added into the fixed cost of the project. Additionally the recovery of costs by year should be added as a revenue stream and discounted. This negatively impacts the NPV of a project when the discounted recovery revenues do not amount to the up-front transmission costs. Transmission costs are a factor in an IPP's decision to construct or not.

9.3 POWER PURCHASE AGREEMENTS (LONG-TERM CONTRACTS)

The investment and development communities have learned a hard lesson from losses on projects built between 2000 and 2003. Currently in the United States, IPPs and their financial backers seek to construct projects with revenue guarantees by contracting with a suitable counterparty. The most common counterparty is a utility. Another potential counterparty may be the ISO. These are the only two counterparties who can pass the risk of long-term contracts through the regulatory arena and ensure cost recovery by a regulated rate mechanism. There are several possibilities for structuring contracts. In this section, we present commonly negotiated contract types and categorize the contract terms. Then we provide a series of issues that the counterparties must agree upon before finalizing contracts. Afterwards we discuss the pricing and valuation of contracts.

9.3.1 Structuring PPAs

The structuring of long-term contracts for the purpose of providing guaranteed revenues to a designated power plant falls into several categories. The categories are defined by dispatchable rights and payment. There are two possibilities for setting the dispatch of a power plant under contract. One is to predefine the dispatch within the contract. For example, the contract may specify that the power plant delivers baseload energy at a profiled output. Below we refer to such contracts as fixed-quantity contracts. The other possibility is for the owner of the plant (usually an IPP) to allow the counterparty (utility) the right to dispatch the plant whenever the counterparty sees fit. This agreement is similar to a leasing agreement: The owner (IPP) leases the plant to the lessee (utility). The owner operates the plant, but has the obligation to dispatch the unit in accordance with the lessee's instructions. Such agreements are referred to as tolling agreements.

There are various ways to structure payments. For fixed-quantity contracts only, the owner may charge the counterparty a fixed payment stream of dollars per MWH. Since the dispatch quantity is fully established within the contract terms, the full payment is also established at the time of the contract's signing. Typically, payments are made on a monthly basis. Most contracts have both fixed and variable components to the payment. The fixed component is referred to as the capacity payment, and the variable component is referred to as the energy payment. Whereas the capacity payment described in the preceding sections is a hypothetical value, in this situation it is an actual contractual payment. Often the contracts are structured so that the energy payment covers the costs of energy production, (fuel, maintenance, and operations), while the capacity payment covers the cost of construction. Below are several ways to structure the energy payment.

- Indexed to power pricc. The energy payment may be set at the settlement price of a designated market. This requires the availability of a price that is published in a forum acceptable to both the owner and the counterparty. An ISO index is an example. Another example is a daily price that is published in an industry-wide accepted publication.
- Indexed to gas price. The energy payment may be indexed to the settlement price of a designated gas market. Once again, an acceptable published price must be available. There is a contractual heat rate that converts the $/MMBTU charge of the gas index into a $/MWH charge for the energy. The owner often sets the contractual heat rate at a level that allows the owner to recover all variable energy costs and is accordingly often higher than the actual operating heat rate of the plant.
- Indexed to coal. Another possibility is for the variable component to be indexed to a different fuel such as coal with a contractual heat rate.
- Determined by guaranteed operational costs. The energy payment may be made in the form of a variable heat rate in accordance with agreed-upon operating characteristics applied to an indexed fuel price. This provides the fuel payment. In addition to the fuel payment there are additional variable operations and maintenance charges along with start-up charges.
- Lessee provides fuel and pays guaranteed operational costs. Under this arrangement the lessee independently arranges for fuel shipments to the plant and pays all of the fuel costs of the dispatch. However, the owner provides a guaranteed efficiency in the form of a variable heat rate. If the actual fuel burn exceeds the contractually guaranteed fuel burn as determined by the contractual heat rate, the owner must reimburse the lessee for the additional fuel cost. Variable operations and maintenance costs along with start-up costs are charged to the lessee at a set cost.
- Lessee provides fuel and pays actual fuel costs along with set VOM and start-up costs. Under this arrangement, the lessee independently arranges for fuel shipments to the plant and pays all of the fuel costs. Additionally, the owner charges variable operations and maintenance costs along with start-up costs.

The terms of the final alternative are very different from those of the other alternatives. In the final alternative, the lessee assumes the risk of the operational efficiency of the plant. This is because the lessee's fuel cost is not guaranteed. With this structure, the owner assumes no financial risk with respect to operational efficiency. Indeed, the payment to the owner is independent of the operational efficiency of the plant.

In all other structures, the owner assumes the risk of the operational efficiency of the plant. The owner may increase earnings by improving operational efficiency or, alternatively, earnings are reduced alongside diminished operational efficiency. Conversely, the lessee's fuel costs are guaranteed by the contract so the lessee does not assume negative risks due to uncertainties in operational efficiency.

Because the lessee assumes the operational efficiency risk in the final alternative, the lessee also must determine the maintenance schedules.

9.3.2 Additional Issues

Aside from determining the dispatchable parameters of the contract along with the form of the payment, there are a host of issues to consider. The issues can be categorized as follows: penalties for operational nonperformance, secondary and market substitution rights, and penalties for financial nonperformance. These issues are discussed below. In the discussion the contract is referred to as the PPA and it is assumed that the owner is an IPP. The counterparty to the IPP is referred to as the counterparty.

9.3.3 Penalties

The counterparty places operational nonperformance penalties in the PPA to encourage the IPP to operate its plant in accordance with performance parameters. There are penalties addressing the following issues.

- Penalty for not meeting construction timeline. The IPP has the obligation to construct a plant and bring it on line by a date specified in the PPA. The PPA also specifies the penalty that the IPP incurs if the on line date is delayed. Normally the penalty is related to the replacement cost of energy and capacity during the delay period.
- Penalty for nondelivery. The PPA specifies an allowable forced outage rate along with a maintenance scheduling protocol. The IPP is subject to a penalty for surpassing the allowable forced outage rate and a disallowance penalty for maintenance outages beyond those specified in the contract. Normally, penalties include the cost of replacement energy during the period of nonperformance as well as foregoing fixed payments that apply to the period of nonperformance. An additional charge is often included to dissuade nonperformance.
- Maximum capacity penalty. The IPP faces an additional penalty for not being able to produce at maximum capacity as specified in the PPA.
- Imbalance penalty. The IPP normally incurs all imbalance costs associated with production that deviates from its scheduled output. In the case that the plant operates in a market without imbalance penalties, an imbalance penalty may be specified in the PPA.
- Ancillary service penalties. The PPA determines the ancillary services that the IPP makes available to the counterparty. Nonperformance penalties are established for agreed-upon ancillary services just as they are for scheduled dispatch.
- Excess fuel burn penalty. In the case where the counterparty provides fuel and the IPP guarantees efficiency through a contractual heat rate, there is a penalty applied to the IPP whenever the fuel burn exceeds that indicated by the guarantee.

9.3.4 Market Substitution

The industry is moving in a direction where a PPA provides the counterparty with all energy and AS benefits. Nevertheless, there have been PPAs structured to allow the IPP market substitution and secondary rights.

- Market substitution rights. The PPA specifies whether or not the IPP has the right to substitute energy output from the designated plant with output from the market. The right allows the IPP to deliver power to the counterparty through a purchase from a third party instead of sourcing it from the designated plant. IPPs will frequently request market substitution rights. This should be viewed with disfavor. Indeed, if the contract is properly structured, market substitution rights should provide no advantage to the IPP. Only when production costs from the designated plant are higher than the market price should there be an advantage in providing market substitution rights. If this occurs, the unit should never dispatch to begin with and there is an underlying problem in the structure of the contract. An example of a structural problem would be a fixed-quantity contract in which associated energy costs are higher than the market price.
- Secondary dispatch and AS rights. For tolling agreements the PPA specifies notification times for scheduling of output. It is the responsibility of the counterparty to inform the IPP of its dispatch decisions within the associated time frame. Most PPAs allow notification during delivery day so that the counterparty can participate in intraday markets. The PPA states whether or not the IPP can dispatch the plant or bid for AS whenever the counterparty forgoes its right by not scheduling within the required time frame.

9.3.5 Default Protection

The PPA also contains provisions that protect both the IPP and the counterparty against default. Defaults may occur for two reasons. One is that the counterparty is in financial stress and unable to perform. The other is that the terms of the PPA become unfavorable to either the IPP or the counterparty and the party viewing the PPA unfavorably wishes to renege on the agreement. The terms may become unfavorable because of changes in the market environment that cause the payment stream to be significantly above or below market conditions. The types of protection in a PPA may include the following.

- Termination rights. The IPP or the counterparty has the right to terminate a contract in the event of a default on the other side's obligations. Should the IPP exercise its right to terminate in the event of a default from the counterparty, the IPP would assume all rights over the capacity and energy benefits. The IPP could then sell these benefits to another party. Conversely, should the counterparty exercise its right to terminate the contract because of nonperformance by the IPP, the counterparty would no longer have the obligation to make payments to the IPP. An example of a default by the IPP would be nonperformance for any of the issues incurring a penalty and subsequent nonpayment of the penalty.

- Collateral. To discourage the IPP or the counterparty from defaulting in the event that the market provides substantially different energy and capacity prices, collateral provisions are contained in the PPA. If the PPA is underpriced relative to the market, the IPP might be tempted to default on the contract and resell capacity and energy benefits in the market at a higher price. Alternatively, if the PPA is overpriced relative to the market, the counterparty might be tempted to default on the PPA and pick up capacity and energy benefits from a cheaper market. The PPA's collateral provisions provide metrics that determine the extent to which the PPA is overpriced or underpriced relative to market conditions. The PPA contains provisions for posting collateral based on the metrics. The party posting collateral would lose its posting if it defaulted.

9.3.6 Pricing and Valuation

Both the IPP and the counterparty must perform a valuation of the terms of a PPA. The IPP determines its pricing in accordance with the valuation, and the counterparty determines whether the offer is acceptable in accordance with its valuation. Here we present a valuation from the perspectives of both the IPP and the counterparty as examples.

EXAMPLE *Capacity Pricing of CCGT and CT Tolling Agreements*

The IPP wishes to use the PPA to back the financing of a power plant. With the framework of discounted cash flows, the present value of the cash flows must be at least equal to the cost of construction for a plant for the project to be of interest. The structure of the energy payment is such that the payment offsets the full variable cost of providing energy. The result is that the present value of cash flows associated with the energy stream (energy payment, start-up costs, VOM) is negligible. Accordingly, the IPP must price the capacity payments at the level where the present value of all capacity payments is at least equal to the cost of construction for the plant. The calculation is carried out below under the following assumptions.

Assumptions

- The contract is for 20 years, and the IPP wants to ensure cost recovery within 20 years.
- The energy payment is structured to offset the variable cost of providing energy; in other words, the present value of the energy payments netted against associated costs is zero.

- The discount rate is 12 percent.
- The capacity payment is a monthly payment that escalates at an annual rate of 2 percent.
- The time of construction for the CT is 2 years. The time of construction for the CCGT is 4 years.

The situation is very much the same as the calculation of the hypothetical capacity payment in Section 9.2. In the example illustrating the calculation of the capacity payment, the value is determined by setting the present value of the capacity payment equal to the cost of construction of a CT. A similar calculation is used here; monthly time frames are used instead of annual time frames.

$$650 = \sum_{j=24}^{482} \frac{CP_j}{1.12^j} = \sum_{j=24}^{482} \frac{CP(1.02)^{j/12}}{1.12^{j/12}}$$

$$850 = \sum_{j=48}^{484} \frac{CP_j}{1.12^j} = \sum_{j=48}^{484} \frac{CP(1.02)^{j/12}}{1.12^{j/12}}$$

where

- The first equation applies to the CT, and the second equation applies to the CCGT.
- CP_j is the capacity payment in \$/kW-month for month j. CP_j escalates at an annual rate of 2% on a monthly basis.
- If one were to assume a 2% annual inflation rate, CP would be the value of each capacity payment in current dollars.
- The full value of the capacity payment for each month is $CP_j \times kW$, where kW is the capacity of the unit in kilowatts. It is common to express the capacity price in \$/kW-month so that comparisons can be made across units with different capacity ratings.

The values of CP are \$6.26/kW-month for the CT and \$9.92/kW-month for the CCGT.

EXAMPLE *Valuation of Tolling Agreements: CT and CCGT*

This example demonstrates the valuation of tolling agreements from the perspective of the counterparty. The assessment is similar to the examples of Section 9.2. The cash flow streams for the assessment include a stream for energy value, a stream for the fuel costs, a stream for start-up costs, a stream for VOM costs, a stream for capacity value, and a stream for the capacity cost. The following assumptions are used.

Assumptions

- The energy and capacity price forecasts from Section 9.2 are used for determining energy and capacity values.
- AS revenues are not included.
- The characteristics of the CCGT and CT are the same as those of the examples in Section 9.2.
- The CCGT is dispatched at maximum output during all on-peak hours.
- For the CCGT, there is 1 week's maintenance outage per year. The maintenance cost is covered by the VOM charge.

- For the CCGT, there is a 2-month overhaul outage every 5 years through year 20. The cost of overhaul is covered by the start-up charge.
- The construction time for the CT is 2 years, and the construction time for the CCGT is 4 years.
- The capacity payment for both the CT and the CCGT is given by the preceding example.
- The tolling agreement is a 20-year agreement.
- The contractual energy payment is equivalent to the energy costs of the examples in Section 9.2.

Case 1. CT

There is no net energy value from the CT tolling agreement. Since the energy payment provides the energy cost and the energy cost of the CT determines the price forecast, the energy payment and energy revenues offset one another, providing a present value of zero.

Similarly, there is no net capacity value from the CT tolling agreement. The discounted cost of construction provides both the forecasted capacity price and the payment. These two streams offset one another, providing a present value for the capacity of zero.

The NPV of the tolling agreement is zero. The interpretation of a zero net value contract is that this is a fair price; the cost of the contract exactly equals the benefits.

Case 2. CCGT

Table 9.3 provides a summary of the revenue streams for the CCGT. For each year of the contract, the energy benefit, fuel costs, start-up costs, VOM, and capacity benefit are identical to those of the CCGT example in Section 9.2. An additional cost associated with the capacity payment is included. Netting all benefits and costs and then discounting back provides the NPV. The resulting NPV is $75 million. The valuation expressed in $/kW-month is $0.69/kW-month. In this case the NPV is positive, indicating that the value from the contract is greater than its costs.

We conclude the chapter with some remarks.

Remarks

- As noted above it is common that a valuation for a CCGT is better than that of a CT. While the NPV is a factor in selecting a contract, there are other factors that must be considered. Companies must select instruments that match the needs of their overall portfolio. Additionally, companies may perform a risk assessment of the various instruments and incorporate the risk assessment into their selection process. The result is that a CT may be selected instead of a CCGT even though the NPV of the CCGT is greater. (Risk assessment is discussed further in Chapter 10.)
- Companies often run competitive solicitations for products such as tolling agreements. The solicitations may target a specific technology such as a CT or a CCGT in line with the needs of the portfolio.
- The valuations in this chapter all use a single discount rate, 12 percent. Some companies may use different discount rates for different projects. The discount

Table 9.3 NPV of CCGT Tolling Agreement

Year	Energy benefit	Fuel cost	Start-up cost	VOM	Capacity benefit	Capacity payment	Future value	Present value
4	135,235,790	(85,135,050)	(6,900,505)	(6,648,840)	36,692,827	(57,983,726)	15,260,497	9,698,321
5	134,455,076	(84,397,950)	(7,038,515)	(6,781,816)	37,426,683	(59,143,400)	14,520,077	8,239,082
6	134,206,308	(84,029,400)	(7,179,285)	(6,917,453)	38,175,217	(60,326,268)	13,929,118	7,056,925
7	130,804,094	(81,449,550)	(7,322,871)	(7,055,802)	38,938,721	(61,532,794)	12,381,798	5,600,897
8	133,420,176	(83,078,541)	(7,469,329)	(7,196,918)	39,717,496	(62,763,450)	12,629,434	5,100,817
9	136,088,579	(84,740,112)	(7,618,715)	(7,340,856)	40,511,845	(64,018,719)	12,882,023	4,645,387
10	138,810,351	(86,434,914)	(7,771,089)	(7,487,673)	41,322,082	(65,299,093)	13,139,664	4,230,620
11	141,586,558	(88,163,612)	(7,926,511)	(7,637,427)	42,148,524	(66,605,075)	13,402,457	3,852,886
12	144,418,289	(89,926,885)	(8,085,041)	(7,790,175)	42,991,494	(67,937,176)	13,670,506	3,508,878
13	147,306,655	(91,725,422)	(8,246,742)	(7,945,979)	43,851,324	(69,295,920)	13,943,916	3,195,586
14	150,252,788	(93,559,931)	(8,411,677)	(8,104,898)	44,728,351	(70,681,838)	14,222,794	2,910,266
15	153,257,844	(95,431,129)	(8,579,911)	(8,266,996)	45,622,918	(72,095,475)	14,507,250	2,650,420
16	156,323,001	(97,339,752)	(8,751,509)	(8,432,336)	46,535,376	(73,537,385)	14,797,395	2,413,776
17	159,449,461	(99,286,547)	(8,926,539)	(8,600,983)	47,466,084	(75,008,132)	15,093,343	2,198,260
18	162,638,450	(101,272,278)	(9,105,070)	(8,773,003)	48,415,405	(76,508,295)	15,395,210	2,001,987
19	165,891,219	(103,297,723)	(9,287,171)	(8,948,463)	49,383,714	(78,038,461)	15,703,114	1,823,238
20	169,209,043	(105,363,678)	(9,472,915)	(9,127,432)	50,371,388	(79,599,230)	16,017,177	1,660,449
21	172,593,224	(107,470,951)	(9,662,373)	(9,309,981)	51,378,816	(81,191,215)	16,337,520	1,512,195
22	176,045,089	(109,620,371)	(9,855,620)	(9,496,180)	52,406,392	(82,815,039)	16,664,270	1,377,177
23	179,565,990	(111,812,778)	(10,052,733)	(9,686,104)	53,454,520	(84,471,340)	16,997,556	1,254,215
							NPV	74,931,380

rate of a riskier project is higher than that of a less risky project. In the above cases an IPP may well apply a higher discount rate when assessing a merchant plant that is not under contract than it would when assessing a plant that has an associated tolling agreement.

Terms of a contract such as penalties, market substitution, and default protection may also influence the choice of a discount rate. The discount rate decreases as the terms become more favorable.

Additionally, the discount rates of the different parties may well differ.

- Within a valuation, it is not uncommon to apply different discount rates to different cash flows depending on the certainty of the cash stream. For example, within a tolling agreement the energy and capacity benefits to the lessee are very uncertain because these benefits are based on forecasts with high uncertainty. However, the capacity payment is fixed by the terms of the contract. It is certain that the purchaser of the contract must pay the capacity payment. The capacity payment stream may be discounted at a significantly lower rate than the other cash flows. This would decrease the NPV of the contract.
- Another valuation methodology that is applied to both tolling agreements and plant valuations parallels that of options theory. The methodology develops a probability distribution of price forecasts as opposed to a single price forecast for gas and power. For each time frame, the future value of the project is then the expected value of the revenue streams with respect to the probability distribution. The NPV is the sum of the discounted future values.
- There are two notes of caution that one should be aware of if one follows an options-type evaluation. The first is that the full options theory that follows the development of Black and Scholes is not applicable. The full options theory applies to liquid market conditions that are not even close to being satisfied in power and gas markets. As such, modifications of the theory are necessary. In particular, the discount rate suggested by the famous Black–Scholes formula is not applicable. Another modification is that one must derive the probability price distributions without market information that is available in transparent and liquid financial markets.

The second note of caution when following an options-type valuation involves the relationship between capacity value and the energy value obtained when electric energy prices are excessive. There is a viewpoint that the capacity value is equivalent to the value of reliability and that this value is expressed when capacity is scarce and there is a corresponding increase in electric energy prices. The level to which prices increase provides a capacity value. An option's valuation captures the value of a plant during extreme energy price periods because some scenarios in the probability distribution contain excessively high prices. Accordingly, the option valuation captures the capacity value as well as the energy value. If one performs an options-type valuation, an additional revenue stream for capacity value should not be considered.

Chapter 10

Risk Management in the Midterm Markets

Midterm planning in a market environment encompasses different objectives than in a traditionally regulated environment. For traditionally regulated utilities, the main objective is to manage an existing portfolio of assets in order to ensure economic availability of physical supply. In a market environment, the objective of a market participant is to align the company's portfolio with the risk preferences of the company. This chapter presents the portfolios of retailers and producers and discusses the use of market instruments to mitigate risk.

The chapter begins with a qualitative discussion of retailer and producer risks in Sections 10.1 and 10.2, respectively. Afterwards, the chapter addresses the quantification of risks. Section 10.3 provides an introduction to the statistical concepts that are used to quantify risk: standard deviation, covariance, and correlation. In Sections 10.4 and 10.5 we quantify the risk of retailer and producer portfolios and demonstrate the impact that market instruments have on retailer and producer portfolio risk. Section 10.6 concludes by comparing risks of integrated electricity suppliers with those of segregated retailers and producers.

10.1 RETAILER RISK

In the financial community the word risk is synonymous with uncertainty. A risk-free contract is one in which there is absolute certainty in the financial outcome. Conversely, a risky contract is one in which there is uncertainty in the financial outcome. The earnings that a retailer earns from contracts with end-use customers depend on many factors including the following:

- Payments from the end customer. Payments depend on the following factors:
  - The customer's contractual price in dollars per kWH (Customer payments are scaled by kWH as opposed to MWH.)
  - The volume of electric energy that the customer consumes

Electric Power Planning for Regulated and Deregulated Markets. By Arthur Mazer

- Procurement costs of the retailer. Procurement costs depend on the following factors:
    - Day-ahead ISO index price
    - Day-ahead purchase volumes
    - Intraday adjustments due to updated load forecasts
    - Imbalance charges
    - Cost of capacity tags
    - Quantity of capacity tags purchased
    - Cost of grid services

There is much uncertainty in both the payment stream from the customer as well as the procurement costs. The actual risk associated with the contract depends on the structuring of the contract. Before discussing possible contract structures, we note that there are two categories for the risk. The first category is price risk. Price risk reflects uncertainty in the ISO index for the delivery dates over the time period of the contract and impacts on the procurement cost to the retailer. There is also price risk associated with the cost of capacity tags, the price of intraday adjustments, and the price of imbalance charges. The second category of risks is volumetric risk. Volumetric risk reflects the uncertainty in customer consumption. Customer consumption is explicit in the payment stream. It is also implicit in most components of the procurement cost stream including the volume that the retailer procures, the day-ahead purchase volumes, intraday purchase adjustments, imbalance fees, and the quantity of capacity tags purchased. Every component associated with a market price or volume is uncertain.

There are two types of volumetric uncertainty that cause procurement cost uncertainty. The first type of volumetric uncertainty is the load forecasting error for a fixed customer base. Forecasting error leads to uncertainties in imbalance penalties and intraday schedule adjustments. The other type of volumetric uncertainty comes about because customers may cancel their contracts with a retailer, effectively setting their purchase volumes to zero.

Retailers structure their risk through the contracts that they sign and the customer base that they target. The risks to the retailer and customer are impacted by the structuring choices. Contracts may be structured to protect the retailer from some risks and pass some risks over to the customer. Conversely, contracts may be structured to protect the customer from risk and maintain risks with the retailer. Contracts protecting the customer from risk are more expensive for the customer than those that pass risks from the retailer to the customer.

As with the structuring of long-term contracts, the retail contracts have several components. There is an energy component, a network component, and a service component. The energy and network components are set to cover energy procurement costs and the cost of networking services. The servicing component allows the retailer to earn a profit.

Below we indicate several possible contract structures for the energy component. The list does not exhaust all possibilities but does give a sense of the types of contracts that exist. The alternatives are ordered by the risk to the retailer, from lowest retailer risk to greatest retailer risk. The associated service component is more expensive for the contracts that are of greater risk to the retailer. All payment alternatives are presented as $/kWH payments.

1. Indexed payment. For all energy that the customer consumes, the customer pays the ISO day-ahead index.
2. Indexed payment with ISO index activated interruptible service. For all energy that the customer consumes, the customer pays the ISO the day-ahead index. Whenever the ISO index is above an agreed-upon threshold, the retailer interrupts the customer's service.
3. Capped index. For all energy that the customer consumes, the customer is charged the ISO day-ahead index as long as the index is below a cap. Otherwise, the customer's payment is equal to the cap.
4. Fixed payment with ISO index-activated interruptible service. For all energy consumed, the customer pays a fixed $/kWH payment. The customer allows the retailer to interrupt service whenever the ISO index is above an agreed-upon threshold.
5. Fixed payment with time-limited interruptible service. For all energy consumed, the customer pays a contractually fixed $/kWH payment. The customer allows the retailer to interrupt service for a specified number of hours per year.
6. Fixed payment. For all energy consumed, the customer pays a contractually fixed $/kWH amount. The customer receives continuous service.

It is of interest to determine the impact of volumetric risk on the overall risk of the contracts. We provide a brief description by considering two cases, one case with fixed volume and the other case with volumetric uncertainty.

CASE 1. *Fixed Volume*

In this case, Structures 1 and 2 eliminate the retailer's price risk and pass the full risk on to the customer. This happens because the retailer schedules the set volume, with interruptible service at a given price through bidding in the day-ahead ISO and the customer pays this full cost. The other structures expose the retailer to varying degrees of price risk.

CASE 2. *Volumetric Uncertainty*

In this case, the retailer is exposed to earnings uncertainty even under Structures 1 and 2. The retailer schedules a volume in accordance with the retailer forecast. The retailer is exposed to intraday prices and the price of the imbalance penalty for the difference between the forecast and the actual consumption of the customer. A point of note that is that volumetric risk leads to price risk. The retailer is always exposed to the price risk for the volume energy associated with forecasting errors.

None of the structures above addresses volumetric risk. For the most part, there is little that can be done to contractually limit volumetric risk. Customers want delivery to meet their requirements, whatever the requirement may be; they do not want to fix the volume. The retailer can address volumetric risk due to forecasting errors by targeting industrial or commercial customers. As noted in Chapter 2, residential load is the most sensitive to daily fluctuations, while industrial load is the least sensitive. Forecasting error of customer load is lower with commercial and industrial customers than it is with residential customers. Industrial customers do receive the most favorable pricing, and one reason is that their load is more predictable.

An issue of great concern for the retailer is the volumetric risk associated with customer departure. Fixed-payment contracts are the riskiest to the retailer. Indeed, a retailer may lose money from a fixed-payment contract if the ISO index becomes expensive. Because of the ISO index risk, a retailer prefers to purchase market instruments, such as a fixed for floating contract, that fix the payment of energy procured on behalf of the customer. Unfortunately, if the customer departs, an instrument purchased to fix the procurement cost of the customer no longer reduces the risk Instead, the retailer becomes exposed to the value of the market instrument and the market value may fall. This is another example of volumetric exposure leading to price exposure. The issue is further addressed in Section 10.4.

Remark

- Another risk to the retailer is the risk that a customer does not make payment. This risk is socialized in a traditional market environment; rates take customer default into account. In a market environment, retailers target customer bases that are less likely to default.

10.2 POWER PRODUCER RISK

The earnings of a producer portfolio are subject to a great deal of uncertainty. The income streams are dependent on many factors including the following:

- Income revenues
  - ISO day-ahead energy index
  - Volume of sales into the ISO energy market
  - ISO ancillary service index
  - Volume of sales into the ISO ancillary service market
  - Intraday market sales
  - Capacity tag sales
- Costs
  - Energy imbalance penalties
  - Capacity unavailability penalties
  - Fuel costs
  - VOM costs

As with the retailer portfolio, the risks fall within two categories, price and volume. The price risk is broader for the producers than it is for retailers. In addition to the price risk in the electricity markets, producers confront price risk in the fuel markets. The volumetric risk is evident in both the incoming revenues and cost streams of a producer's portfolio.

The volumetric risk is related to forced outages. Producers have some control over this risk. They can adopt operational and maintenance policies that are in line with risk preferences. For example, a producer can choose to push equipment to operational limits and forgo discretionary maintenance. Such a policy may save expenses and allow a unit to operate during a profitable period. However, this policy increases the risk of forced outages and their associated imbalance penalties and capacity unavailability penalties. The alternative is to exercise greater caution in the use of a unit and follow a more disciplined maintenance regime. This policy may force the producer to bring a unit offline during profitable hours when the producer would prefer to keep the unit in the market. However, over the long run the unit operates more predictably, securing more certain revenues.

Aside from the maintenance policies of units, another way for a producer to reduce the volumetric uncertainty is through the ownership of many plants. Statistical reasoning demonstrates that the percentage of availability of a portfolio of plants is more certain than the percentage of availability for fewer plants. This idea is pursued in Section 10.5.

A producer would like to use market instruments to reduce price risk. For example, a producer might sell a fixed for floating contract that addresses the price uncertainty of the ISO energy index. The situation is similar to the retailer that wants to fix the cost of procuring on behalf of a customer through the purchase of a fixed for floating contract. The retailer is concerned about volumetric risk due to customer departure and has no control over this risk. On the other hand, through proper maintenance policies and ownership of a large portfolio, a producer can manage volumetric risk. The producer is accordingly more willing to use market instruments for risk mitigation. The issue is further explored in Section 10.5.

10.3 A QUICK RISK PRIMER IN STATISTICS FOR RISK MANAGEMENT

This section provides a brief primer in basic statistical concepts that are useful in financial risk management. The section begins with definitions that are used to measure risk and provides a formulaic basis for quantifying the risk of a portfolio of assets. In the remainder of the chapter, the formulas are applied to the portfolio of retailers and producers.

Recall that risk is synonymous with uncertainty. Using historical data, market information and modeling assumptions, potential financial outcomes of a portfolio are assigned probabilities. Analysts consider a statistical measurement known as the standard deviation of the outcomes as a measurement of risk because the standard deviation of a set of outcomes provides a measurement of uncertainty of the outcomes. The formalism behind this concept is presented along with some examples. We first present algebraic definitions of the concepts. Afterwards, geometric interpretations are given.

Definition: Discrete Probability Set. A discrete probability set is a discrete set of outcomes with a probability assigned to each outcome.

EXAMPLE *A Fair Coin*

The outcome set is {H, T}. The probability of each outcome is 1/2.

EXAMPLE *A Single Die*

The outcome set is {1, 2, 3, 4, 5, 6}. The probability of each outcome is 1/6.

EXAMPLE *A Set of Dice*

The outcome set is {2, 3, 4, 5, 6, 7, 8, 9, 10, 11, 12}. The probability of the outcomes differs for each outcome. The probability is as follows.

$P(2) = 1/36$ $P(3) = 2/36$ $P(4) = 3/36$ $P(5) = 4/36$ $P(6) = 5/36$ $P(7) = 6/36$

$P(8) = 5/36$ $P(9) = 4/36$ $P(10) = 3/36$ $P(11) = 2/36$ $P(12) = 1/36$

where $P(s)$ is the probability that the outcome is s.

Note that the probabilities are all positive and probabilities over all outcomes sum to 1, that is, $P(2) + P(3) + P(4) + \ldots + P(12) = 1$. These properties provide the definition of a discrete probability distribution.

Definition: Discrete Random Variable. A discrete random variable is a real valued function over the outcomes of a discrete probability set. Often, a random variable is represented by a capital letter.

EXAMPLE *Different Outcomes for a Penny Toss*

$X = -1$ if a penny toss lands tails.
$X = 1$ if a penny toss lands heads.

EXAMPLE *Same Outcome for a Penny Toss*

$X = 0$ if a penny toss lands either heads or tails.

EXAMPLE *Outcome Given by the Roll of a Die*

$X =$ the roll of a die.

EXAMPLE ***Outcome Given as a Function of the Roll of a Die***

X = twice the role of a die.

EXAMPLE ***Outcome Given as the Roll of Two Dice***

X = the roll of two dice.

Definition: Expected Value. The expected value of a discrete random variable, X, is given by the formula:

$$E(X) = \sum_{j=1}^{N} X_j P_j$$

where X_j is the value associated with outcome j and P_j is the probability of outcome j.

The expected value is also known as the mean or the average value. $E(X)$ is read as the expected value of X.

EXAMPLE ***Different Outcomes for a Penny Toss***

$X = -1$ if a penny toss lands tails.
$X = 1$ if a penny toss lands heads.

$$E(X) = -\frac{1}{2} + \frac{1}{2} = 0$$

EXAMPLE ***Same Outcome for a Penny Toss***

$X = 0$ if a penny toss lands either heads or tails.

$$E(X) = 0$$

EXAMPLE ***Outcome Given by the Roll of a Die***

X = the roll of a die.

$$E(X) = \frac{1}{6} + \frac{2}{6} + \frac{3}{6} + \frac{4}{6} + \frac{5}{6} + \frac{6}{6} = \frac{42}{6} = 3.5$$

EXAMPLE ***Outcome Given as a Function of the Roll of a Die***

X = twice the role of a die.

$$E(X) = \frac{2}{6} + \frac{4}{6} + \frac{6}{6} + \frac{8}{6} + \frac{10}{6} + \frac{12}{6} = \frac{84}{6} = 7$$

EXAMPLE ***Outcome Given as the Roll of Two Dice***

X = the roll of two dice.

$$E(X) = 2\left(\frac{1}{36}\right) + 3\left(\frac{2}{36}\right) + 4\left(\frac{3}{36}\right) + 5\left(\frac{4}{36}\right) + 6\left(\frac{5}{36}\right) + 7\left(\frac{6}{36}\right)$$
$$+ 8\left(\frac{5}{36}\right) + 9\left(\frac{4}{36}\right) + 10\left(\frac{3}{36}\right) + 11\left(\frac{2}{36}\right) + 12\left(\frac{1}{36}\right) = 7$$

We next move toward quantification of uncertainty. A random variable is most certain when all outcomes lead to the same result, which is the expected value. The example in which all penny tosses lead to the assigned value zero is such a case. Any metric quantifying uncertainty that is applied to a constant value should yield a value of zero because there is zero uncertainty in the constant value.

In general, the uncertainty is low when the random variable has a high probability of being close to its expected value. Conversely, uncertainty is high when a random variable has a high probability of being distant from its expected value. Indeed, if we are highly certain of an outcome, knowing its expected value provides a lot of information: The outcome should be close to the expected value. However, if there is a lot of uncertainty, knowing the expected value provides little information: The outcome may be far from the expected value. The following definitions capture this property.

Definition: Variance. The variance of a random variable X is given by the following formula:

$$\text{Var}(X) = \sum_{j=1}^{N} (E(X) - X_j)^2 P_j$$

Note that all the terms in the sum of the variance are positive numbers. It follows that the variance is always positive.

Definition: Standard Deviation. The standard deviation of a random variable X is given by the following formula:

$$\text{SD}(X) = \sqrt{\text{Var}(X)}$$

EXAMPLE ***Different Outcomes for a Penny Toss***

$X = -1$ if a penny toss lands tails.
$X = 1$ if a penny toss lands heads.

$$E(X) = -\frac{1}{2} + \frac{1}{2} = 0$$

$$\text{Var}(X) = (0-(-1))^2\left(\frac{1}{2}\right) + (0-1)^2\left(\frac{1}{2}\right) = 1$$

$$\text{SD}(X) = \sqrt{1} = 1$$

EXAMPLE ***Same Outcome for a Penny Toss***

$X = 0$ if a penny toss lands either heads or tails.

$$E(X) = 0$$
$$\text{Var}(X) = (0-0)^2(1) = 0$$
$$\text{SD}(X) = 0$$

EXAMPLE ***Outcome Given by the Roll of a Die***

$X =$ the roll of a die.

$$E(X) = \frac{1}{6} + \frac{2}{6} + \frac{3}{6} + \frac{4}{6} + \frac{5}{6} + \frac{6}{6} = 3.5$$

$$\text{Var}(X) = \frac{(3.5-1)^2}{6} + \frac{(3.5-2)^2}{6} + \frac{(3.5-3)^2}{6} + \frac{(3.5-4)^2}{6} + \frac{(3.5-5)^2}{6} + \frac{(3.5-6)^2}{6} = 2.917$$

$$\text{SD}(X) = \sqrt{2.917} = 1.708$$

EXAMPLE ***X = Twice the Role of a Die***

$$E(X) = \frac{2}{6} + \frac{4}{6} + \frac{6}{6} + \frac{8}{6} + \frac{10}{6} + \frac{12}{6} = 7$$

$$\text{Var}(X) = \frac{(7-2)^2}{6} + \frac{(7-4)^2}{6} + \frac{(7-6)^2}{6} + \frac{(7-8)^2}{6} + \frac{(7-10)^2}{6} + \frac{(7-12)^2}{6} = 11.67$$

$$\text{SD}(X) = \sqrt{11.67} = 3.416$$

EXAMPLE ***X = the Roll of Two Dice***

$$E(X) = \frac{2}{36} + \frac{3}{36} + \frac{4}{36} + \frac{5}{36} + \frac{6}{36} + \frac{7}{36} + \frac{8}{36} + \frac{9}{36} + \frac{10}{36} + \frac{11}{36} + \frac{12}{36} = 7$$

$$\text{Var}(X) = \frac{(7-2)^2}{36} + \frac{(7-3)^2 2}{36} + \frac{(7-4)^2 3}{36} + \frac{(7-5)^2 4}{36} + \frac{(7-6)^2 5}{36} + \frac{(7-7)^2 6}{36}$$
$$+ \frac{(7-8)^2 5}{36} + \frac{(7-9)^2 4}{36} + \frac{(7-10)^2 3}{36} + \frac{(7-11)^2 2}{36} + \frac{(7-12)^2}{36} = 5.833$$

$$\text{SD}(X) = \sqrt{5.833} = 2.415$$

Note that the standard deviation has the properties desired in a measurement of uncertainty. The standard deviation of the coin toss with a certain outcome, $X = 0$ for every outcome, is zero. Also note that the standard deviation of twice the roll of a single die is higher than that of the roll of two dice, 3.416 compared to 2.415. The expected value for both of these cases is 7. Figures 10.1 and 10.2 provide plots of the discrete probability distributions for the two cases. The plots are bar graphs in which the height of each bar is the probability of the assigned outcome. Note that the probability distribution associated with the roll of two dice is clustered toward

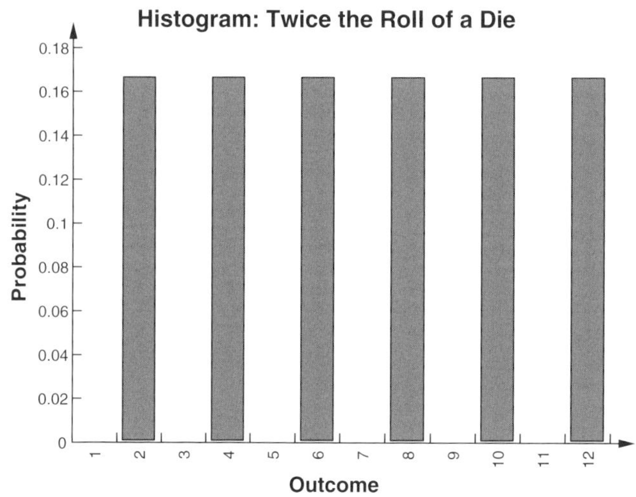

Figure 10.1

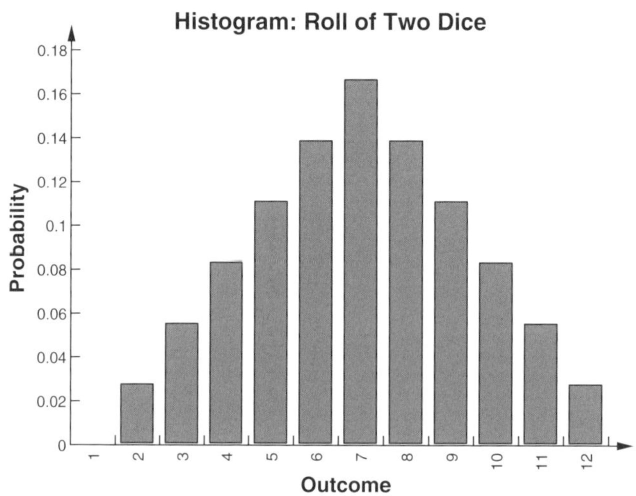

Figure 10.2

the expected value in comparison with the probability distribution associated with twice the roll of one die. This is as it should be; probability distributions of random variables with higher uncertainty are spread out, whereas probability distributions of random variables with lower uncertainty are clustered toward the expected value.

There is another formula for the variance and a corresponding formula for the standard deviation that is equivalent to the above definitions.

$$\text{Var}(X) = E(X^2) - [E(X)]^2$$

$$\text{SD}(X) = \sqrt{E(X^2) - [E(X)]^2}$$

These formulas often provide a more convenient way to calculate the variance or standard deviation.

In finance theory, the earnings from a single asset is considered a random variable, and the earnings of a portfolio is a combination of random variables. Our goal is to quantify the risk of a portfolio of energy assets for both retailers and producers. To accomplish this we need to develop a formulaic approach for measuring the standard deviation of a combination of assets. The first step is an understanding of the joint probability of two outcomes. We demonstrate this with an example.

EXAMPLE *Toss of a Penny and a Dime*

There are four elements in the set of joint outcomes {(H,H) (H,T) (T,H) (T,T)}. The probability of each outcome is 1/4.

Let $X = 1$ whenever the penny is heads.
$= -1$ whenever the penny is tails.

Let $Y = 1$ whenever the dime is heads.
$= -1$ whenever the dime is tails.

Determine $E(X + Y)$.

Table 10.1 shows all the outcomes for X and Y. From the table, one calculates $E(X + Y)$ as follows:

$$E(X+Y) = 2\left(\frac{1}{4}\right) + 0\left(\frac{1}{4}\right) + 0\left(\frac{1}{4}\right) - 2\left(\frac{1}{4}\right) = 0$$

Table 10.1 Joint Probability Distribution

Penny	Dime	X	Y	$X + Y$	P
H	H	1	1	2	.25
H	T	1	−1	0	.25
T	H	−1	1	0	.25
T	T	−1	−1	−2	.25

Note that because all the outcomes have the same probability, it is possible to determine $E(X + Y)$ by averaging over all the values in the last column of the table.

Definition: Covariance. The covariance between two random variables is given by the following formula:

$$\mathrm{Cov}(X,Y) = E((X - \bar{X})(Y - \bar{Y}))$$

$$\bar{X} = E(X)$$

$$\bar{Y} = E(Y)$$

One point to note is that the expectation of the first equation defining the covariance is with respect to the joint outcomes of two random variables, X and Y. To determine the covariance requires the joint set of outcomes along with a joint probability distribution as in the preceding example.

The covariance of two random variables satisfies the following inequalities and motivates another definition.

$$-\mathrm{SD}(X)\mathrm{SD}(Y) \leq \mathrm{Cov}(X,Y) \leq \mathrm{SD}(X)\mathrm{SD}(Y)$$

Definition: Correlation. The correlation between two random variables is given by the following formula:

$$\mathrm{Corr}(X,Y) = \frac{\mathrm{Cov}(X,Y)}{\mathrm{SD}(X)\mathrm{SD}(Y)}$$

The correlation provides a measurement of how well the relation between two random variables can be estimated by a line. If the relationship between two random variables is precisely described by a line with positive slope, their correlation is 1. Alternatively, if the relationship between two random variables is precisely described by a line with negative slope, their correlation is negative 1. Correlations other than 0 indicate that there is a line that provides some information on the relationship between the variables. The estimation of the relation between two random variables by a well-chosen line improves as the correlation approaches 1 or negative 1. Two random variables that are totally unrelated have correlation 0. Additionally, for two random variables that are related, but the relationship is not at all linear, the correlation is 0. Random variables that increase in synchrony have positive correlation. Random variables that are in asynchrony (one decreases as the other increases) have negative correlation. Some examples provide motivation for and understanding of this concept. We first start with qualitative examples and then move to quantitative examples. Finally, a geometric interpretation is presented.

EXAMPLE *Positive Correlation*

Let X = the price of a raw material and Y = the price of a finished product composed of Y. X and Y are positively correlated ($\mathrm{Corr}(X,Y) > 0$) since the cost of the raw material increases the cost of the finished product. One could think of X as the price of natural gas and Y as the price of electric power.

EXAMPLE *Negative Correlation*

Let X = the price of a home and Y = the number of people who can afford the home. X and Y are negatively correlated (Corr(X,Y) < 0). As the price of the home increases, the number of people who can afford it decreases.

EXAMPLE *Zero Correlation*

Let X = the temperature in Sydney, Australia and Y = the winning time in a leg of the Tour de France. These variables have correlation 0. If one were to gather the temperatures in Sydney on the days of Tour de France races and compare the temperatures with the winning times, there would be no relationship between the times and the temperatures.

Below are examples that demonstrate the calculation of the covariance and correlation.

EXAMPLE *Calculation of Covariance and Correlation: Unrelated Variables*

$X = -1$ if a penny toss lands tails.
$X = 1$ if a penny toss lands heads.
Y = the roll of a die.

The outcomes and results necessary to calculate the correlation are provided in Table 10.2. Each line in the table represents a joint outcome, and the probability of each joint outcome is 1/12.

Since each joint outcome has the same probability, the covariance is obtained by averaging the values in the last column. For this example, both Cov(X,Y) and Corr(X,Y) are 0. These two random variables are completely unrelated; the outcome of the flip of the penny has no impact whatsoever on the outcome of the toss of the die, and vice versa.

Table 10.2 Outcomes of Coin Toss and Die Roll

Coin	Die	X	Y	$X - E(X)$	$Y - E(Y)$	$(X - E(X))*(Y - E(Y))$
T	1	−1	1	−1	−2.5	2.5
T	2	−1	2	−1	−1.5	1.5
T	3	−1	3	−1	−0.5	0.5
T	4	−1	4	−1	0.5	−0.5
T	5	−1	5	−1	1.5	−1.5
T	6	−1	6	−1	2.5	−2.5
H	1	1	1	1	−2.5	−2.5
H	2	1	2	1	−1.5	−1.5
H	3	1	3	1	−0.5	−0.5
H	4	1	4	1	0.5	0.5
H	5	1	5	1	1.5	1.5
H	6	1	6	1	2.5	2.5

EXAMPLE *Calculation of Covariance and Correlation: Related Variables*

X = twice the roll of a die
Y = the role of a pair of dice where the first die is the same as used to determine X.

Table 10.3 provides outcomes and results necessary to calculate the covariance and correlation. Each line represents a joint outcome, and the probability of each joint outcome is 1/36.

Table 10.3 Outcomes for Roll of Two Dice

Die 1	Die 2	X	Y	$X - E(X)$	$Y - E(Y)$	$(X - E(X))*(Y - E(Y))$
1	1	2	2	−5	−5	25
1	2	2	3	−5	−4	20
1	3	2	4	−5	−3	15
1	4	2	5	−5	−2	10
1	5	2	6	−5	−1	5
1	6	2	7	−5	0	0
2	1	4	3	−3	−4	12
2	2	4	4	−3	−3	9
2	3	4	5	−3	−2	6
2	4	4	6	−3	−1	3
2	5	4	7	−3	0	0
2	6	4	8	−3	1	−3
3	1	6	4	−1	−3	3
3	2	6	5	−1	−2	2
3	3	6	6	−1	−1	1
3	4	6	7	−1	0	0
3	5	6	8	−1	1	−1
3	6	6	9	−1	2	−2
4	1	8	5	1	−2	−2
4	2	8	6	1	−1	−1
4	3	8	7	1	0	0
4	4	8	8	1	1	1
4	5	8	9	1	2	2
4	6	8	10	1	3	3
5	1	10	6	3	−1	−3
5	2	10	7	3	0	0
5	3	10	8	3	1	3
5	4	10	9	3	2	6
5	5	10	10	3	3	9
5	6	12	12	5	5	25
6	1	12	7	5	0	0
6	2	12	8	5	1	5
6	3	12	9	5	2	10
6	4	12	10	5	3	15
6	5	12	11	5	4	20
6	6	12	12	5	5	25

The results for the covariance and correlations are as follows:

$$\text{Cov}(X,Y) = 5.833$$

$$\text{SD}(X)\text{SD}(Y) = 8.25$$

$$\text{Corr}(X,Y) = .707$$

Because the probabilities of the joint outcomes are the same, the covariance is obtained by finding the average of the values in the last column.

10.3.1 Scatter Plots

Scatter plots provide a series of outcomes between two random variables on planar coordinates. Below are examples of scatter plots that indicate the data for highly correlated and poorly correlated variables. The horizontal axis gives the outcome of the random variable *X*, while the vertical axis gives the outcome of random variable *Y*. Analysts create scatter plots to analyze actual data. For example, the *X* variable may represent the price of crude oil while the *Y* variable represents the earnings of the oil industry.

EXAMPLE ***Two Variables with Positive Correlation***

The scatter plot of Figure 10.3 presents a sample of two random variables with a correlation of .85. The provides a good estimate of the outcomes of the variables.

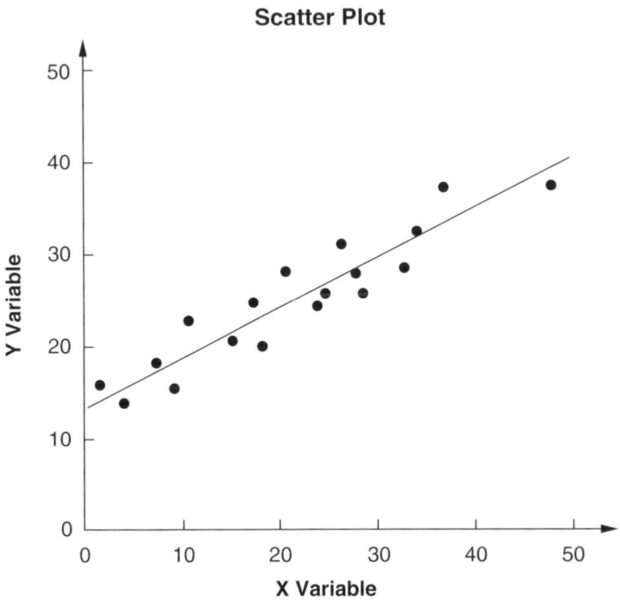

Figure 10.3

EXAMPLE *Two Variables with Negative Correlation*

The scatter plot of Figure 10.4 presents a sample of two random variables with a correlation of –.85. The fit of the line to the data is as good as the fit of the previous example with positive correlation.

EXAMPLE *Two Totally Unrelated Variables*

The scatter plot of Figure 10.5 shows a sample of two random variables with zero correlation. These variables have no apparent relation between them; indeed, they are independent variables. There is no line that provides an estimate of the outcomes.

EXAMPLE *Two Related Variables with Zero Correlation*

The final scatter plot of Figure 10.6 shows a sample of two random variables that also have zero correlation. Although one can discern a definite relationship between the variables that is evidenced by its clear parabola-like outline, the relationship is not well described by a line.

We next provide formulas that are useful when calculating the risk of portfolios.

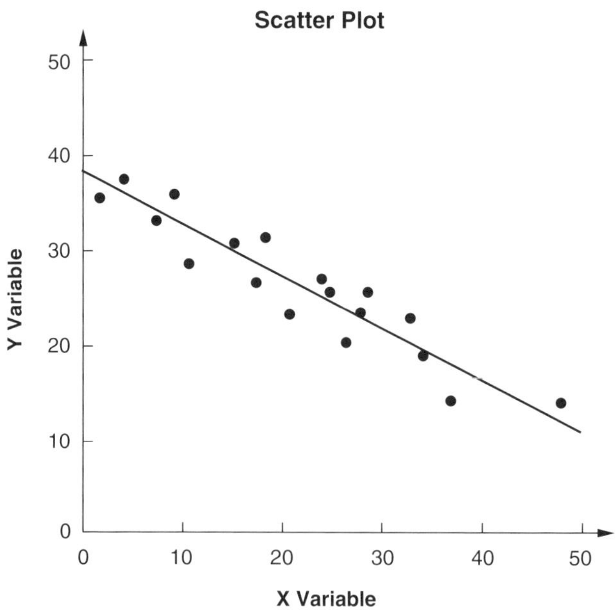

Figure 10.4

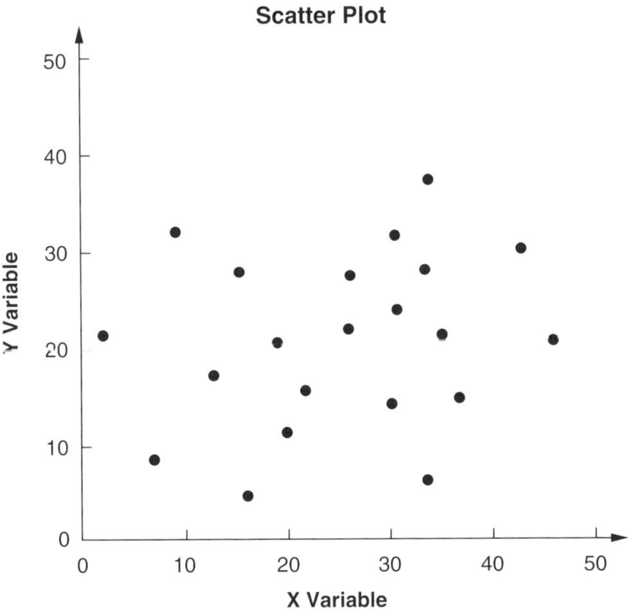

Figure 10.5

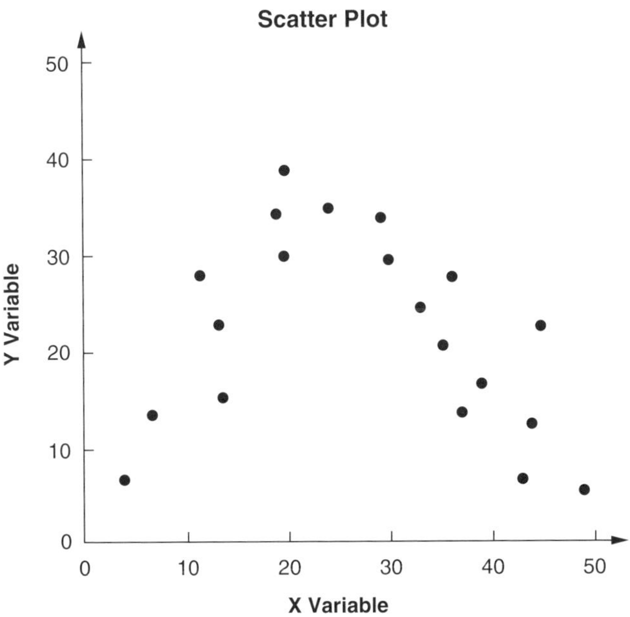

Figure 10.6

Formula 1

$$\begin{aligned}\mathrm{Var}(X + Y) &= \mathrm{Var}(X) + \mathrm{Var}(Y) + 2\mathrm{Cov}(X,Y) \\ &= \mathrm{Var}(X) + \mathrm{Var}(Y) + 2\mathrm{Corr}(X,Y)\mathrm{SD}(X)\mathrm{SD}(Y)\end{aligned}$$

Formula 2

$$\mathrm{Var}(bX) = b^2\mathrm{Var}(X)$$

whenever b is a constant number.

Formula 3

$$\mathrm{Cov}(aX,Y) = a\mathrm{Cov}(X,Y)$$

$$\mathrm{Cov}(X,bY) = b\mathrm{Cov}(X,Y)$$

$$\mathrm{Cov}(aX,bY) = ab\mathrm{Cov}(X,Y)$$

whenever a and b are constants.

In the case that X and Y are independent, that is, totally unrelated, the following formula holds.

Formula 4

$$\mathrm{Var}(XY) = \mathrm{Var}(X)\mathrm{Var}(Y) + [E(X)]^2\mathrm{Var}(Y) + [E(Y)]^2\mathrm{Var}(X)$$

In the case that the $\mathrm{Corr}(X,Y) = 1$, the following formula holds.

Formula 5

$$\mathrm{Var}(XY) = \frac{\mathrm{Var}(Y)\mathrm{Var}(X^2)}{\mathrm{Var}(X)} = \frac{\mathrm{Var}(X)\mathrm{Var}(Y^2)}{\mathrm{Var}(Y)}$$

We note that for any two random variables X and Y, we can write the following:

Formula 6

$$Y = aX + Z$$

where

$$a = \frac{\mathrm{Cov}(X,Y)}{\mathrm{Var}(X)} = \mathrm{Corr}(X,Y)\frac{\mathrm{SD}(Y)}{\mathrm{SD}(X)}$$

and

$$\mathrm{Corr}(X,Z) = 0$$

This allows one to decompose Y into a component that is correlated with X and a component that is uncorrelated with X. This decomposition is frequently used.

The above formula has an extension to more variables.

Formula 7

$$W = aX + bY + Z$$

with

$$\text{Corr}(X,Z) = \text{Corr}(Y,Z) = 0$$

We close this section with some final remarks.

- We selected discrete random variables to minimize technical requirements and simplify the exposition. All the definitions and formulas apply to continuous random variables as well. A bit more sophistication is required to carry out computations.
- For discrete random variables, Corr(X,Y) = 1 is identical to the relation $Y = aX + b$, where a and b are constants. For those familiar with regression, a and b are regression coefficients and there is no error term.
- For the more general case, $Y = aX + Z$, a is a regression coefficient and Z is a random variable with nonzero mean. We can make this appear more like a standard regression, $Y = aX + b + Z'$ with $Z = b + Z'$. Both a and b are regression coefficients and Z' is a random variable with zero mean.

10.4 RISK MANAGEMENT IN MIDTERM MARKETS: RETAILERS

In this section, we assess the risk of the retailer portfolio and illustrate the use of market instruments to reduce risk. The section begins with a description of the retailer's portfolio as a set of random variables. Afterward, a method for determining the risk of the portfolio and selection of market instruments is presented. Some examples illustrate the risks that retailers confront and the difficulties in mitigating these risks. Additional attention is given to the problem of risk mitigation with uncertainty in the size of the retailer's customer base due to customer switching to another retail service provider.

10.4.1 Description of a Retailer's Portfolio

A retailer's portfolio consists of a set of customer contracts that obligate the retailer to deliver power throughout a given time frame in the future. The central issue of this section is to find the quantity of market instruments that most effectively mitigates the risk associated with signing up customers. We consider the risk of a portfolio for the delivery period of a single month. This is relevant because midterm markets trade in monthly time blocks. Additionally, we assume that the only contracts between the retailer and the customer base are fixed-price contracts and that the price is the same for all contracts.

A retailer's portfolio is structured ahead of delivery month with the anticipation of earnings at the end of delivery month. The words portfolio and earnings are used synonymously, and a description of the portfolio is given by a description of its earnings. The equation for a retailer's anticipated earnings on signing up a customer base without the introduction of market instruments is the following:

$$R = L(p - n - P) - D - I - K$$

All of the lower-case letters are fixed numbers.

- Lower-case p is the fixed price of the contracts between the retailer and the customer base.
- Lower-case n is the fixed network fee in \$/MWH that the ISO charges for its grid services.

All of the capitalized variables are random variables.

- R is the composite value of the retail position in dollars over the delivery month.
- L is the monthly load obligation of the customer base in MWH.
- P is the load-weighted day-ahead ISO index in dollars per MWH.
- D is the intraday costs of adjusting the portfolio in dollars.
- I is the imbalance penalty in dollars.
- K is the cost of capacity meeting the ISO capacity tag requirement. Note that K depends on price and the peak hour load, both of which are uncertain.

Again we emphasize that none of these variables is known on signing the contract. The retailer can only estimate probability distributions for these variables.

By load-weighted day-ahead ISO index, we mean that each hourly ISO day-ahead settlement price is weighted by the percentage of load demand for that hour and then the result is averaged.

$$P = \sum_{h=1}^{m} \frac{L_h P_h}{L}$$

- m is the total number of hours in the month
- L_h is the load for hour h
- P_h is the ISO index for hour h

An example illustrates this concept.

EXAMPLE *Calculation of P*

Consider the case in which the total load for the month is 10 MWH and all this load is consumed during only 2 hours of the month, Hour 1 and Hour 30. (Note that in a 30-day month there are 720 hours.) Assume that 2 MWH is consumed in Hour 1 and 8 MWH is consumed in Hour 30. Let P_1 indicate the day-ahead ISO price for Hour 1 and P_{30} indicate the day-ahead ISO price for Hour 30. Then $P = .2P_1 + .8P_{30}$. Note that LP indicates the total day ahead energy cost to the retailer.

The set of risks inherent in the retailer portfolio are given by the following terms.

1. Energy revenues risk, *LP*
2. Day-ahead energy cost risk, *LP*
3. Intraday price adjustment risk, *LD*
4. Imbalance penalty risk, *LI*
5. Capacity penalty risk, *K*
6. ISO grid cost risk, *LN*

Note that volumetric risk is present in each of these factors either through the load requirement, *L*, or the capacity term, *K*.

The risk in this portfolio is the standard deviation. For convenience, we compute the variance with the understanding that the risk is the square root of the variance. Using Formulas 1, 2, and 3 from Section 10.3, we obtain the following expression for the variance.

$$\begin{aligned}\mathrm{Var}(R) = {} & (p-n)^2\mathrm{Var}(L) + \mathrm{Var}(LP) + \mathrm{Var}(D) + \mathrm{Var}(I) + \mathrm{Var}(K) - 2(p-n)(\mathrm{Cov}(L,LP) \\ & - 2\mathrm{Cov}(L,D) - 2\mathrm{Cov}(L,I)) - 2(p-n)\mathrm{Cov}(L,K) + 2\mathrm{Cov}(LP,D) \\ & + 2\mathrm{Cov}(LP,I) + 2\mathrm{Cov}(LP,K) + 2\mathrm{Cov}(D,I) + 2\mathrm{Cov}(D,K) + 2\mathrm{Cov}(I,K)\end{aligned}$$

This expression is difficult to analyze without making simplifying assumptions. In this section we make simplifying assumptions that allow for a perspective on individual risk components. Cases are developed that allow us to focus on price risk and load risk separately In all of these cases the objective is to minimize the risk of a portfolio with market instruments. The use of market instruments to minimize risk is called hedging.

The concept of risk mitigation is depicted in the following graphs. The first graph in Figure 10.7 depicts the discrete probabilities of a given retailer portfolio,

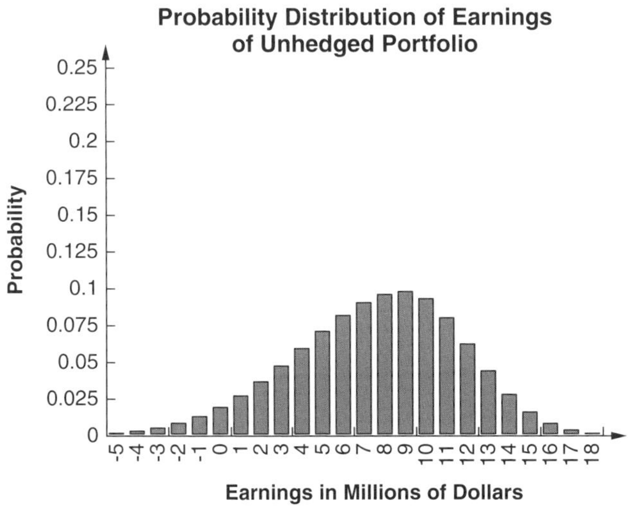

Figure 10.7

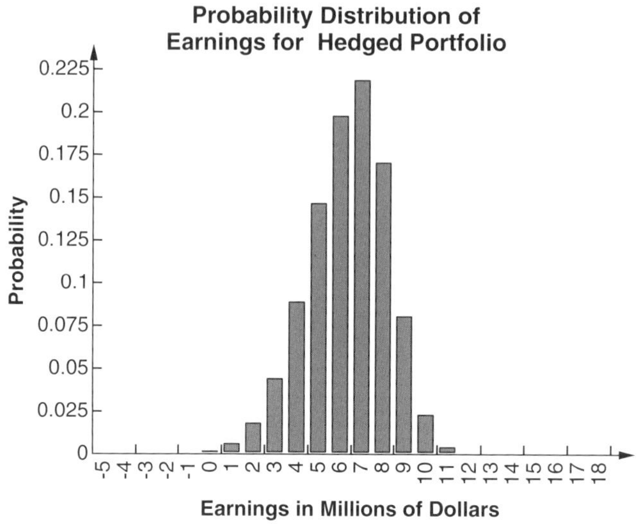

Figure 10.8

R. The second graph in Figure 10.8 depicts the discrete probabilities of the same portfolio with risk-mitigating market instruments included.

There are a couple of points to remark upon before continuing.

Remarks

- The purpose of hedging is to reduce uncertainty. Uncertainty of the hedged portfolio is less than that of the unhedged portfolio because outcomes are more clustered about the mean. The unhedged portfolio has a wider distribution.
- Hedging generally reduces the expected value of the portfolio. In the graph in Figure 10.7 the expected value of the unhedged portfolio is \$7.55 million, while the expected value of the hedged portfolio, Figure 10.8, is \$6.34 million. The difference in expected values is a price that is paid for reducing the uncertainty.

10.4.2 Method for Selecting Market Instruments that Minimize Risk

Below is a step-by-step method for selecting the market instruments that minimize the risk of the portfolio. The selection is referred to as the optimal hedge. This method is used repeatedly throughout the remainder of the chapter.

Step 1. Develop a set of simplifying assumptions.

As noted above, the expression Var(R) is complicated. This expression becomes more difficult to analyze with the addition of market instruments in the portfolio. Simplifying the expression is necessary for analysis of results.

Step 2. Determine the set of market instruments that are incorporated in the portfolio.

Rather than incorporating all possible market instruments, it is often instructive to initially reduce the set so that the portfolio does not become too complicated for analysis. In the following examples we introduce one market instrument.

Step 3. Write down the expression for the variance of the portfolio with market instruments included. Develop further simplifying assumptions.

Once the variance of the portfolio with market instruments is known, it may be useful to simplify even further. Below we eliminate any terms that are unaffected by the addition of market instruments. Also in the examples below, this expression is quadratic in the term that represents the quantity of the market instrument to be bought or sold. This means that the variance is a parabola of the following form:

$$\text{Variance}(e) = A_2 e^2 + 2A_1 e + A_0$$

- e is the decision variable, the quantity of the market instrument bought or sold.
- A_j are the quadratic coefficients.

Step 4. Solve for the quantity of the market instruments that minimizes the risk.

There are several ways to do this. One can use the methods of calculus to determine minimums. In this chapter, because the variance is a parabola with respect to the decision variable, e, one can solve for the vertex of the parabola. The solution is the following:

$$e^* = -\frac{A_1}{A_2}$$

The asterisk indicates that the solution is the risk-minimizing solution. This convention is used throughout the remainder of the text.

Step 5. Determine the value of the risk associated with the optimal hedge.

Once the optimal hedge is known, the variance and then the standard deviation of the minimal risk portfolio can be determined. In our case the value becomes the following:

$$\begin{aligned}\text{Variance}(e^*) &= A_2\left(\frac{A_1}{A_2}\right)^2 - 2A_1\left(\frac{A_1}{A_2}\right) + A_0 \\ &= A_0 - \frac{A_1^2}{A_2}\end{aligned}$$

$$\text{SD}(e^*) = \sqrt{A_0 - \frac{A_1^2}{A_2}}$$

EXAMPLE 1. *Contractually Fixed Load*

In this example we apply the five-step method to the retailer's portfolio, assuming that the load is set at a contractual volume.

Step 1. Develop a set of simplifying assumptions.

The first set of assumptions is the following.

- L and K are constant.
- No correlation among all the variables; accordingly, all the covariance terms disappear.

With these assumptions, the expression for Var(R) simplifies to the following.

$$\mathrm{Var}(R) = L^2\mathrm{Var}(P) + \mathrm{Var}(D) + \mathrm{Var}(I)$$

Step 2. Determine the set of market instruments that are incorporated in the portfolio.

Recall the set of market instruments available to the retailer; this is given below.

1. Fixed for floating contract
2. Monthly indexed call option
3. Daily indexed call option
4. Monthly indexed put option
5. Daily indexed put option

There are no instruments that address the risks inherent in intraday schedule adjustment costs or imbalance penalties, D or I, as the financial settlement of the instruments does not depend on the outcomes in the intraday market or the imbalance penalty. The financial settlement of these instruments depends on the ISO day-ahead index. Accordingly, the instruments may be used to mitigate the risk of the first term, P. Below, we illustrate that it is not possible to completely mitigate the payment risk in the first term even under the assumption that the load is constant. To simplify the discussion, only the fixed for floating contract is considered.

Recall that fixed for floating contracts are available in block times, on-peak period, off-peak period, and baseload (round the clock). Consider the baseload contract. Let B denote the average baseload price over the delivery month; B is a random variable at the time the contract is set. Let f be the fixed price of the fixed for floating contract. Both B and f are in \$/MWH. Then the value per MWH to the holder of the contract is $B - f$.

There is a critical point to note. B and P are related, but they differ. They are both weighted averages of the ISO index over the delivery month; however, the weightings differ. The weighting of B is uniform for each hour of the delivery month, whereas the weighting for P is set by the load shape of the customer base. This is a fundamental problem. Retailers cannot completely mitigate the day-ahead energy costs of their customer base with a baseload profiled market instrument unless their customer load profiles are baseload. Recall the load shapes in Chapter 2. Typical profiles are not baseload.

While we illustrate this problem using the baseload instrument, the problem remains for any instrument or combination of instruments in the broker market. The customer load profile generally differs from the profile of available instruments, so the available instruments cannot completely mitigate price risk.

Still, the retailers wish to somewhat mitigate their risk. Assume that the retailer purchases e MWH of the contract; e is the energy volume of the contract. Then the retailer's portfolio is the following:

$$\begin{aligned} R + e(B - f) &= L(p - P - n) - D - I - K + e(B - f) \\ &= eB - LP + L(p - n) - D - I - K - ef \end{aligned}$$

Step 3. Write down the expression for the variance of the portfolio with market instruments included and develop further simplifying assumptions.

Assume that B has zero correlation with all variables except P. Using Formulas 1, 2, and 3 form Section 10.3 we calculate the variance of this portfolio as follows:

$$\begin{aligned} \mathrm{Var}(R + e(B - f)) &= \mathrm{Var}(R) + e^2\mathrm{Var}(B - f) + 2e\mathrm{Cov}(R,B - f) \\ &= \mathrm{Var}(R) + e^2\mathrm{Var}(B) + 2e\mathrm{Cov}(R,B) \\ &= L^2\mathrm{Var}(P) + \mathrm{Var}(D) + \mathrm{Var}(I)) + e^2\mathrm{Var}(B) + 2e\mathrm{Cov}(L(p - P - n) - K, B) \\ &= L^2(\mathrm{Var}(P) + \mathrm{Var}(D) + \mathrm{Var}(I)) + e^2\mathrm{Var}(B) - 2eL\mathrm{Cov}(P,B) \\ &= L^2(\mathrm{Var}(P)) + e^2(\mathrm{Var}(B)) - 2eL\mathrm{Cov}(P,B) + (\mathrm{Var}(D) + \mathrm{Var}(I)) \end{aligned}$$

The final terms involving D and I represent risks in intraday markets and imbalance penalties. These risks cannot be addressed using available market instruments. We subsequently focus on the first three terms and ignore the last two.

Rewriting the first three terms and rearranging gives the result as a quadratic function in e.

$$\begin{aligned} \mathrm{Var}(R + e(B - f)) &= L^2(\mathrm{Var}(P)) + e^2(\mathrm{Var}(B)) - 2eL\mathrm{Cov}(P,B) \\ &= \mathrm{Var}(B)e^2 - 2\mathrm{Cov}(P,B)Le + L^2(\mathrm{Var}(P)) \end{aligned}$$

The last equality emphasizes that the result is a quadratic function in e; the variance is a parabola with respect to the value e. One could mitigate the price risk by setting the expression to 0. Using the quadratic formula we solve for e.

$$\begin{aligned} e &= \frac{L\mathrm{Cov}(P,B) \pm \sqrt{(L\mathrm{Cov}(P,B))^2 - L^2\mathrm{Var}(P)\mathrm{Var}(B)}}{\mathrm{Var}(B)} \\ &= \frac{L\mathrm{Cov}(P,B) \pm \sqrt{(L\mathrm{Corr}(P,B))^2\mathrm{Var}(P)\mathrm{Var}(B) - L^2\mathrm{Var}(P)\mathrm{Var}(B)}}{\mathrm{Var}(B)} \\ &= \frac{L\mathrm{Cov}(P,B) \pm \sqrt{(\mathrm{Corr}(P,B)^2 - 1)L^2\mathrm{Var}(P)\mathrm{Var}(B)}}{\mathrm{Var}(B)} \end{aligned}$$

There is a real solution only when the term inside the square root is not negative. This occurs only when $\mathrm{Corr}(P,B) = 1$. In this case, B and P are identical, $B = P$, and the above equation reduces to $e = L$. The retailer would purchase L MW of baseload; retailers would be able to eliminate their day-ahead energy cost risk.

For the general case, B and P differ because the load profile of the customer base differs from the baseload profile of the market instrument.

Step 4. Solve for the quantity of the market instruments that minimizes the risk.

The value of e that minimizes the variance is the value associated with the vertex of the graph of the variance. In general, the vertex of the equation

$$\text{Variance}(e) = A_2e^2 + 2A_1e + A_0$$

is given at the point

$$e^* = -\frac{A_1}{A_2}$$

In the above case, $A_1 = -\text{Cov}(P,B)L$ while $A_2 = \text{Var}(B)$.

This gives the following optimal value, e^*.

$$\begin{aligned} e^* &= \frac{\text{Cov}(P, B)\text{L}}{\text{Var}(B)} \\ &= \text{Corr}(P, B)L\frac{\text{SD}(P)}{\text{SD}(B)} \end{aligned}$$

Step 5. Determine the value of the risk associated with the optimal hedge.

Ignoring the terms with D and I, and setting e to its optimal value, the price risk becomes

$$\begin{aligned} \text{SD}(R + e^*(B - f)) &= \sqrt{\text{Var}(B)e^2 - 2\text{Cov}(P,B)Le + L^2(\text{Var}(P))} \\ &= \sqrt{\frac{\text{Cov}(P,B)^2L^2 - 2\text{Cov}(P,B)^2L^2}{\text{Var}(B)} + L^2(\text{Var}(P))} \\ &= \sqrt{\frac{L^2\text{Var}(P) - \text{Cov}(P,B)^2L^2}{\text{Var}(B)}} \\ &= L\sqrt{\text{Var}(P) - \text{Corr}(P,B)^2\text{Var}(P)} \\ &= \sqrt{1 - \text{Corr}(P,B)^2}\,L\text{SD}(P) \end{aligned}$$

We present a few remarks before discussing the case with load uncertainty.

Remarks

- It is possible to improve price risk mitigation by using a combination of on-peak, off-peak, and baseload products to improve the match between the customer load profile and the market instrument profile.
- In a liquid market, the value f is quoted and provides a market view of the value of power.
- Note that the assumption above is that the total load for delivery month is known; however, there is no assumption about load shape. Uncertainty in the load shape provides additional uncertainty to the price, P.
- Another complication that is not discussed is the fact that a day-ahead ISO index is set for every load bus; however, markets do not trade at every bus. Markets may trade on a basket of load buses or a subset of dominant buses. Therefore, the load of the day-ahead ISO index that applies to the retailer purchases from the ISO may differ from the settlement price with the financial counterparty.

- Another point to note is that the load, L, refers to the load at the load bus; not the load delivered to the customer. The fixed price, p, is discounted to reflect the fact that the end customer pays for a lower volume of energy.

EXAMPLE 2. *Load Uncertainty with Fixed Index*

This example applies the five-step approach for risk mitigation to the case in which load is uncertain and price is fixed. We show that the fixed for floating market instrument is unable to hedge any risk in this portfolio.

Step 1. Develop a set of simplifying assumptions.

- The customer base has signed contracts with fixed payment, p.
- We ignore the capacity payment.
- The variables D and I are neglected.
- $P = B = f$. The index is the same as the baseload contract. Both are fixed at the forward market value.

Step 2. Determine the set of market instruments that are incorporated in the portfolio.

The only market contract considered is the baseload fixed for floating.

Step 3. Write down the expression for the variance of the portfolio with market instruments included and develop further simplifying assumptions.

Neglecting capacity payments, intraday risk, and imbalance penalties, the expression for the future earnings of the retailer's portfolio is given as follows:

$$R = L(p - n - P)$$

Calculating the variance of this portfolio gives the following.

$$\text{Var}(R) = (p - n - P)^2\text{Var}(L)$$

As before, we examine portfolio with the addition of fixed for floating contracts and try to determine the optimal hedging volume, e.

$$\text{Var}(R + e(B - f)) = \text{Var}(R) + \text{Var}(B - f)e^2 + 2\text{Cov}(B - f, R)e = \text{Var}(R)$$

Since the only random variable is R, all other terms vanish.

Step 4. Solve for the quantity of the market instruments that minimizes the risk.

The portfolio risk is independent of the choice of e; the value remains the same no matter what value of e is selected.

Step 5. Determine the value of the risk associated with the optimal hedge.

$$\text{SD}((R + e(B - f)) = \text{SD}(R) = (p - n - P)\text{SD}(L)$$

The risk in this portfolio depends on the standard deviation of the load, and one is unable to mitigate this risk with a fixed for floating instrument.

EXAMPLE 3. *Load Uncertainty with Price Uncertainty*

This example applies the approach of the preceding example to the case in which load and price are both uncertain. The outcome for the minimal risk solution depends on the correlation between price and revenue.

Step 1. Develop a set of simplifying assumptions.

- The customer base has signed contracts with fixed payment, p.
- We ignore the capacity payment.
- The variables D and I are independent of all other variables; all covariance terms involving these variables are 0.
- $\text{Corr}(B,P) = 1$, which implies $P = B$.

Note that the final assumption is very strong. It states that there is a market instrument that is a perfect match for the weighted day-ahead ISO index of the retailer's customer base. Even with this strong assumption, we show that the retailer must assume risk.

Step 2. Determine the set of market instruments that are incorporated in the portfolio.

The only market contract considered is the baseload fixed for floating.

Step 3. Write down the expression for the variance of the portfolio with market instruments included and develop further simplifying assumptions.

Recall the expression for the future value of the retailer's portfolio with $K = 0$.

$$R = L(p - P - n) - D - I$$

As in the previous case where we considered load fixed, the terms D and I are independent of all other terms as well as market instruments. It is not possible to mitigate the intraday price risk or the imbalance penalty. As such, we will exclude these risks from the portfolio solely for the purpose of simplifying expressions. We examine the first term.

$$R = L(p - n - P)$$

Calculating the variance of this portfolio gives the following.

$$\text{Var}(R) = (p - n)^2\text{Var}(L) + \text{Var}(LP) - 2(p - n)\text{Cov}(L,LP)$$

As before, we examine portfolio with the addition of fixed for floating contracts and try to determine the optimal hedging volume, e.

$$\begin{aligned}\text{Var}(R + e(B - f)) &= \text{Var}(R) + e^2\text{Var}(B - f) + 2e\text{Cov}(R,B - f)\\ &= \text{Var}(R) + e^2\text{Var}(B) + 2e\text{Cov}(R,B)\\ &= \text{Var}(P)e^2 + 2\text{Cov}(R,P)e + \text{Var}(R)\\ &= \text{Var}(P)e^2 + 2\text{Cov}(L(p - n - P),P)e + \text{Var}(L(p - n - P))\end{aligned}$$

Note that, consistent with our assumptions, we replaced B with P. The first term represents the variance in the ISO index. The second term represents the covariance between the cash flow of delivered power and the ISO index, while the final term is the variance in the cash flow of delivered power. Note that the expression is quadratic in e; the portfolio variance is a parabola with respect to the value e.

Next, we divert momentarily from the steps for selecting e^* and determine whether it is possible to eliminate risk. Elimination of risk requires setting the above expression to 0. As before, we not that the variance is quadratic in e and use the quadratic formula to set the variance to 0 by solving for e.

$$e = \frac{-\text{Cov}(L(p-n-P),P) \pm \sqrt{\text{Cov}(L(p-n-P),P)^2 - \text{Var}(L(p-\text{n}-P))\text{Var}(P)}}{\text{Var}(P)}$$

$$= \frac{-\text{Cov}(L(p-n-P),P) \pm \sqrt{\begin{matrix}\text{Corr}\left(L((p-n-P),P)^2\text{Var}(L(p-n-P),P)\right)\text{Var}(P) \\ -\text{Var}(L(p-n-P))\text{Var}(P)\end{matrix}}}{\text{Var}(P)}$$

$$= \frac{\left(-\text{Corr}(L(p-n-P),P) \pm \sqrt{\text{Corr}(L(p-n-P),P)^2 - 1}\right)\text{SD}((L(p-n-P))\,\text{SD}(P)}{\text{Var(P)}}$$

$$= \left(-\text{Corr}(L(p-n-P),P) \pm \sqrt{\text{Corr}(L(p-n-P),P)^2 - 1}\right)\frac{\text{SD}((L(p-n-P))}{\text{SD}(P)}$$

$$= \left(-\text{Corr}(R,P) \pm \sqrt{\text{Corr}(R,P)^2 - 1}\right)\frac{\text{SD}(R)}{\text{SD}(P)}$$

A value of e that mitigates risk is possible only if the term in the square root is nonnegative. This occurs only if the cash flow, $L(p-n-P)$, and the ISO index, P, are fully correlated or fully uncorrelated. Under all other circumstances, it is not possible to mitigate the risk.

Step 4. Solve for the quantity of the market instruments that minimizes the risk.

As before, the value of e that provides the minimum risk is found by determining the value of e where the vertex of the parabola gives the following:

$$e^* = -\frac{\text{Cov}(P,L(p-n-P))}{\text{Var}(P)}$$

$$= -(\text{Corr}(P,L(p-n-P)))\frac{\text{SD}(L(p-n-P))}{\text{SD}(P)}$$

$$= -\text{Corr}(P,R)\frac{\text{SD}(R)}{\text{SD}(P)}$$

Recall that P is the ISO day-ahead index and the term $L(p-n-P)$ is the cash flow of delivered power excluding intraday adjustments and imbalance fees. The minimal risk value, e^*, is proportional to the ratio of the risk of the cash flow of delivered power over price index risk. The term in front of this ratio is the negative of the correlation between price and earnings. If the correlation is negative, higher P results in lower earnings and e^* is positive. In this case, one purchases the fixed for floating contract. Alternatively, if the correlation is positive, higher P results in higher earnings and e^* is negative. In this case, one sells the fixed for floating contract.

Step 5. Determine the value of the risk associated with the optimal hedge.

The risk associated with this choice for e is the following.

$$\text{SD}((R + e^*(B-f)) = \sqrt{\text{Var}(P)e^{*2} + 2\text{Cov}(L(p-n-P),P)e^* + \text{Var}(L(p-n-P))}$$

$$= \sqrt{\frac{\text{Var}(P)(\text{Cov}(P,L((p-n)-P))}{\text{Var}(P)^2} - \frac{2\text{Cov}(L(p-n-P),P)^2}{\text{Var}(P)} + \text{Var}(L(p-n-P))}$$

$$= \sqrt{\text{Var}(L(p-n-P)) - \frac{\text{Cov}(L(p-n-P),P)^2}{\text{Var}(P)}}$$

$$= \sqrt{\text{Var}(L(p-n-P)) - \text{Corr}(L(p-n-P),P)^2\text{Var}(L(p-n-P))}$$

$$= \sqrt{1 - \text{Corr}(L(p-n-P),P)^2}\,\text{SD}(L(p-n-P))$$

$$= \sqrt{1 - \text{Corr}(R,P)^2}\,\text{SD}(R)$$

This expression informs us that the minimum risk is a multiple of the unhedged risk. The multiple is given by the coefficient, $\sqrt{1-\text{Corr}(P,R)^2}$. Consistent with the finding at the end of Step 3, risk may be mitigated provided that there is high positive or negative correlation between the portfolio cash flow and the index. This is further discussed below.

A Geometric Interpretation of Example 3

This subsection introduces a geometric interpretation of hedging that is useful. The material is motivated by the previous example and provides an explanation of the result.

The above examples demonstrate the difficulties associated with mitigating combined load and price uncertainty. The result, e^*, of Example 3 is particularly noteworthy. The result states that the risk mitigation solution is to either buy or sell the fixed for floating contracts depending on the correlation between the ISO index and the portfolio cash flows on delivery, Corr(R,P). The portfolio of Example 3 contains price risk because there is uncertainty in the ISO index. Without the use of any market instruments the volume associated with this price risk is L. With the use of market instruments the volume associated with the price risk is $L - e$. It is somewhat surprising that one risk mitigation solution increases the volumetric exposure. This happens whenever e^* is negative, or equivalently whenever Corr(R,P) is positive. In that case, the solution contributes to the portfolio volume rather than offsetting the load. One would think that this increases the associated risk, when in fact it decreases the risk. Alternatively, e^* is positive whenever Corr(R,P) is negative. We discuss this outcome and provide evidence that the correlation is near zero. In such a case, it is not possible for the retailer to mitigate risk with the fixed for floating instrument. This issue is further explored.

This section adopts the following two assumptions.

- $R = L(p - n - P)$; we ignore other terms.
- $B = P$

The complete portfolio with the sale of a fixed for floating contract is the following:

$$\begin{aligned} R + e(B - f) &= R + e(P - f) \\ &= L(p - n - P) + e(P - f) \\ &= L(p - n) - ef - (L - e)P \end{aligned}$$

Indeed, the coefficient to the term with price risk is $L - e$. This can be considered the volume associated with the price risk.

First we demonstrate the correctness of the result from Example 3. In particular, we show that increasing the volumetric exposure does indeed make sense whenever Corr(R,P) is positive. Recall that two random variables are positively correlated whenever their tendency is to have common directional movements; both tend to move upward and downward together. The graph in Figure 10.9 is a plot of sample outcomes for R versus P in the case of positively correlated R and P.

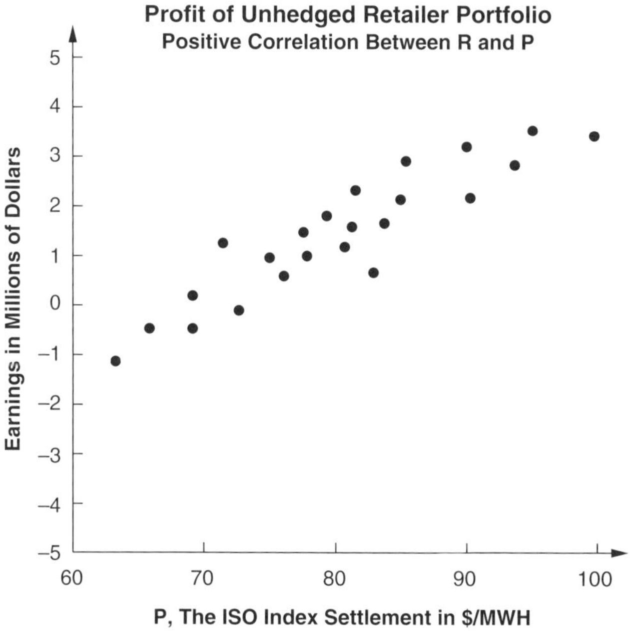

Figure 10.9

The graph in Figure 10.10 provides the best linear approximation for the profit/ ISO index relationship. The linear approximation can be found using the regression formula, Formula 5 of Section 10.3.

A key geometric interpretation of the optimal hedge is that it is the hedge that completely hedges the linear approximation. The slope of the approximation is identically the negative of e^*. The slope of the above approximation is 35,000 MWH/ dollar. Setting e^* to −35,000 MWH, that is, selling the fixed for floating at 35,000 MWH provides the optimal hedge. The graph in Figure 10.11 is a plot of outcomes for the optimal hedge. The vertical axis gives the earnings, and the horizontal axis gives the ISO index. Note that the earnings are a function of the index: Once the index is known the earnings are specified. Also, since e is negative, $e =$ −35,000 MWH, the earnings decrease with increasing P.

Combining the retailer portfolio with the sale of the fixed for floating contract provides a hedged portfolio. The earnings on the hedged portfolio are the sum of two individual earnings streams. Adding the results of the fixed for floating sale onto the data points from the retailer portfolio provides a plot of earnings for the hedged portfolio as illustrated in Figure 10.12. Comparing this graph with that of the unhedged portfolio we note that the vertical spread of the outcomes is reduced. Note that for low values of P, high earnings from the sale of fixed for floating instrument offset the losses from the unhedged retail portfolio. Similarly, for high values of P, the high earnings of the unhedged portfolio offset losses from the sale of the fixed for floating instrument. While the earnings outcomes of the unhedged portfolio range

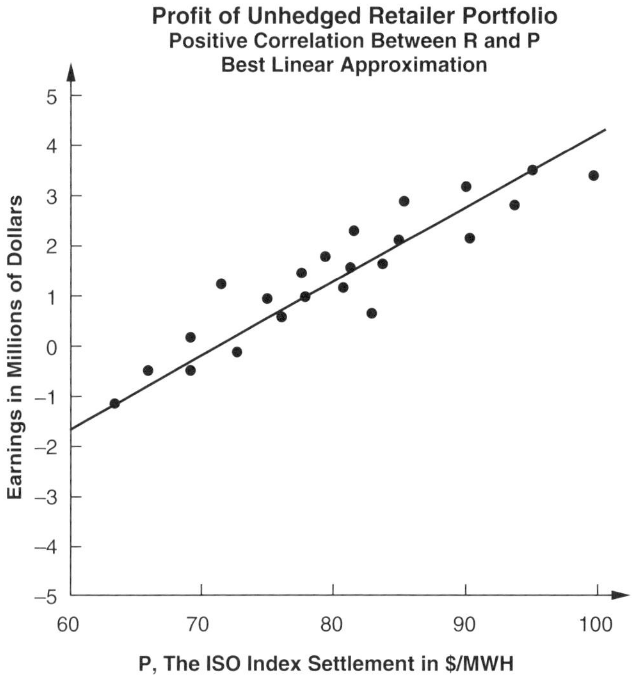

Figure 10.10

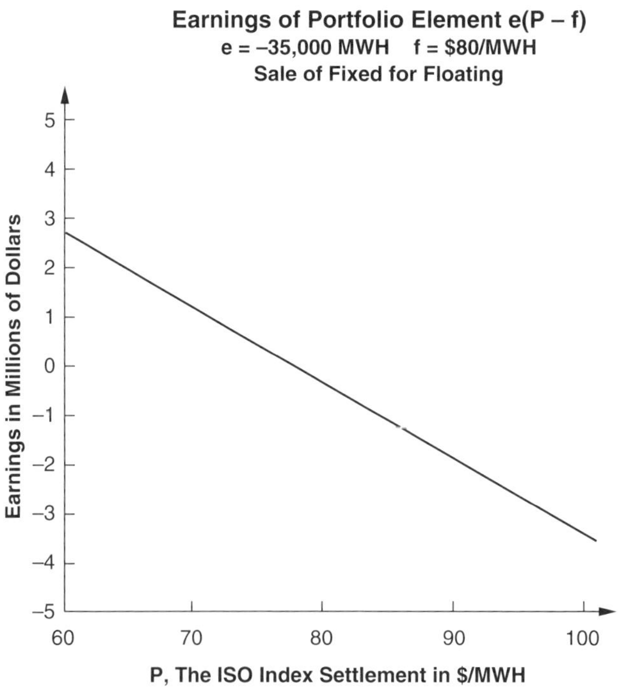

Figure 10.11

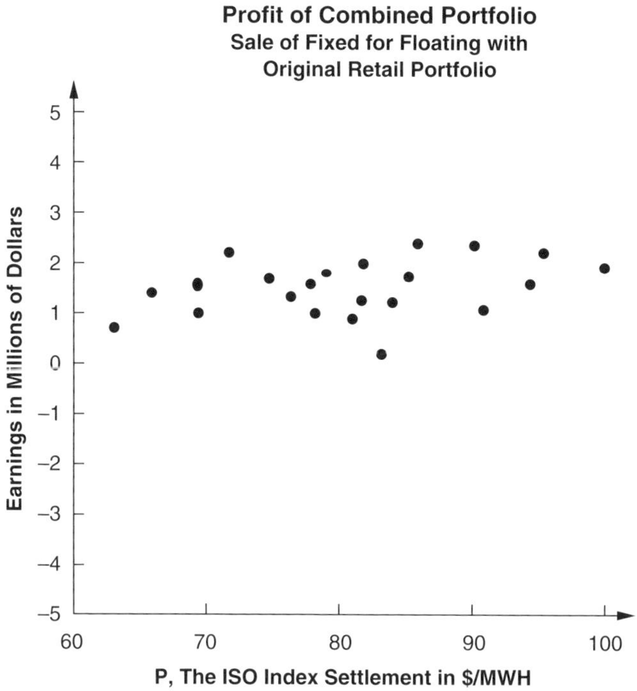

Figure 10.12

from −1.1 million to 3.3 million dollars, the earnings outcomes of the hedged portfolio only range between 0 and 2.6 million dollars.

One obtains the expected result; there is more certainty in the outcomes of the hedged portfolio than the unhedged portfolio as outcomes of the hedged portfolio are clustered in a tighter range than the outcomes in the unhedged portfolio. It is possible to perform a similar analysis for the case when R and P are negatively correlated. While the slopes of the first two graphs would be interchanged, their outcomes would be offsetting.

We make one further note. If one were to add the results of the optimal hedge with the linear approximation, the result would be a constant output at around \$1.3 million.

The reader may have noted a disturbing feature in the above analysis. The first graph is not realistic. It is not possible for the profit to increase with increasing ISO index. In fact, the opposite seems more intuitive since when P rises above the net of the sales price and network costs, the expression $L(p - n - P)$ becomes negative and the portfolio accrues losses for each sale that it incurs. On the other side, it is possible that cash flows decrease as the ISO index decreases because of decreased L; as the ISO index price falls, customers depart from the retailer, looking for better prices. We further discuss the market mechanisms that guide customer and competitive behavior and then investigate the correlation, Corr(R,P).

There is a market view of the expected settlement price for the ISO index. This is the fixed cost of a fixed for floating instrument. We call this fixed cost the forward price, and in the upcoming discussion we consider the forward price to be an indicator of the ISO index. The forward price changes in response to activities of market participants. Retailers attract new customers with pricing that is dependent on the forward price. When the forward price falls, competitive retailers attract customers by offering lower prices. A retailer with a customer base pegged to a higher forward price loses customers. When prices rise, a retailer's customers do not depart. We next demonstrate how this dynamic is modeled and how it affects the correlation, Corr(R,P).

There are two preliminary concepts to introduce, the retailer's demand function and profit function. For a given product, the demand function provides a retailer with the number of units of the product that the retailer can sell at a given price. An example of a retailer's demand function is presented in the graph in Figure 10.13.

As expected, the quantity of product that the retailer sells increases as the price decreases. We provide a slight variation of the demand function. Rather than considering the demand to be a function of price alone, we consider it to be a function of the profit margin. This is the net cash flow between the customer and ISO, ($p - n - P$). The relationship between the profit margin and the demand is the same as that of the conventional demand function: Demand decreases with increasing profit margin.

A critical decision for the retailer is the profit margin that the retailer demands. If the retailer is aware of the demand function, it is possible to multiply the volume by the profit margin and arrive at the total profit. Once total profit is known, the price, p, can be set at the point that maximizes the profit. In our case, the demand is expressed by load. We illustrate this with a linear demand function.

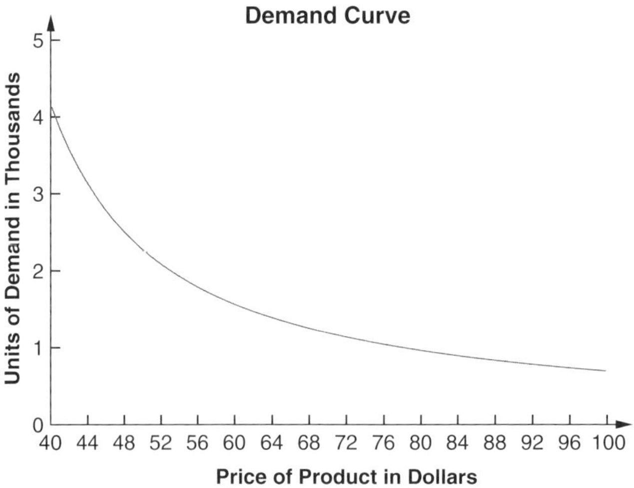

Figure 10.13

EXAMPLE 4. Profit Function with Linear Demand

Let the demand function be given by the following.

$$L = -a(p - n - P) + b$$

Both a and b are positive values. The profit is the volume times the profit margin.

$$\begin{aligned}\text{Profit} &= L(p - n - P) \\ &= -a(p - n - P)^2 + b(p - n - P)\end{aligned}$$

The profit is quadratic in the profit margin; the graph is a parabola sloping downward.

The profit margin that maximizes profit is given by the vertex.

$$p - n - P = \frac{b}{2a}$$

$$p = \frac{b}{2a} + n + P$$

Once the profit-maximizing solution is identified, the retailer sets the price. The relation between profit and forward price (ISO index) is given in the graph in Figure 10.14.

In light of the example, we can see that the scatter plot with positive correlation between R and P is not realistic. A more realistic plot is given in Figure 10.15. The outcomes are low when the forward price and corresponding ISO index fall because customers depart; competitive retailers are able to provide better pricing. There is a range of ISO index outcomes where customers remain with the retailer and profit margins are attractive. However, once the ISO index gets beyond the customer price and grid charges, the retailer begins to lose money. Losses can be substantial. In this case, the correlation between earnings and the ISO index is positive in the region to

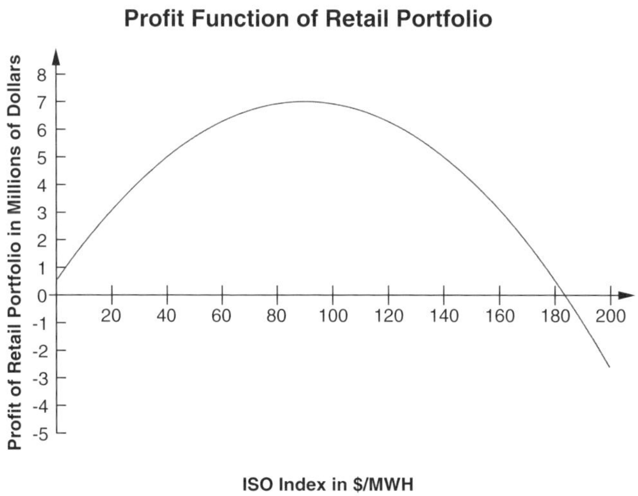

Figure 10.14

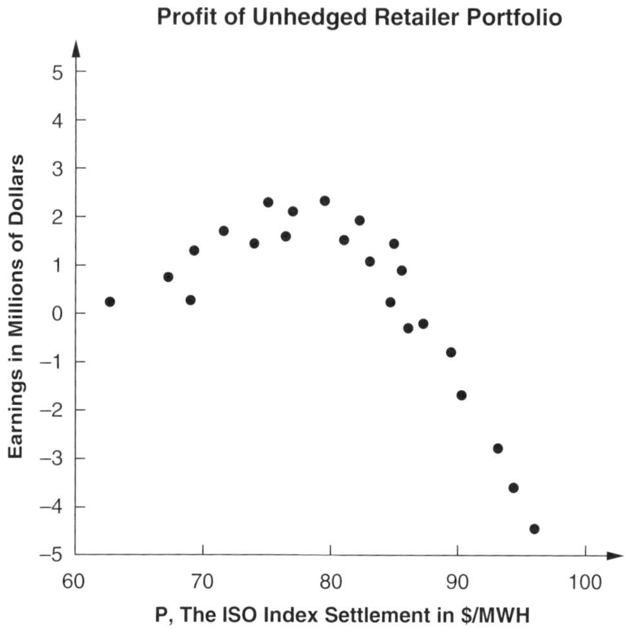

Figure 10.15

the left of the peak, where the graph is increasing, and negative in the region to the right of the peak, where the graph is decreasing. It is not evident whether the retailer should hedge by purchasing or selling the fixed for floating instrument.

Perhaps one might be worried by losses that one would incur if the ISO index increases. In that case, one would purchase a fixed for floating instrument. The earnings from the instrument are given in the graph in Figure 10.16.

Summing the earnings from the graphs provides the profit for the combined portfolio. The purchase of the fixed for floating instrument does stem losses that occur when the ISO index is high. However, when the forward prices fall and the ISO index follows, the fixed for floating instrument does not hedge the retail sales. Instead, profits erode into losses when the outcome of the ISO index is low. The range of profit outcomes of the original portfolio is between –4.6 million and 2.3 million dollars. With the addition of the fixed for floating instrument the range of outcomes is between –5 million and 3.6 million dollars. The instrument increases the span of outcomes, increasing the overall risk. The plot of Figure 10.17 illustrates the result.

Instead of purchasing the fixed for floating instrument, one could sell the fixed for floating instrument. This would mitigate profit erosion when the ISO index is low, but would exacerbate losses when the ISO index is high.

We present some remarks before continuing.

Remarks

- A glance at the relation between profit and ISO index indicates that there is no good linear approximation to the outcomes. Because there is no good linear

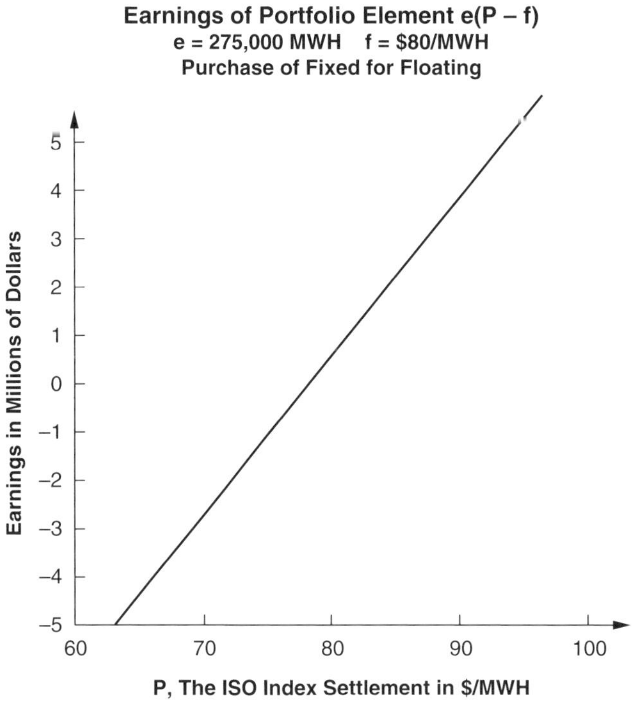

Figure 10.16

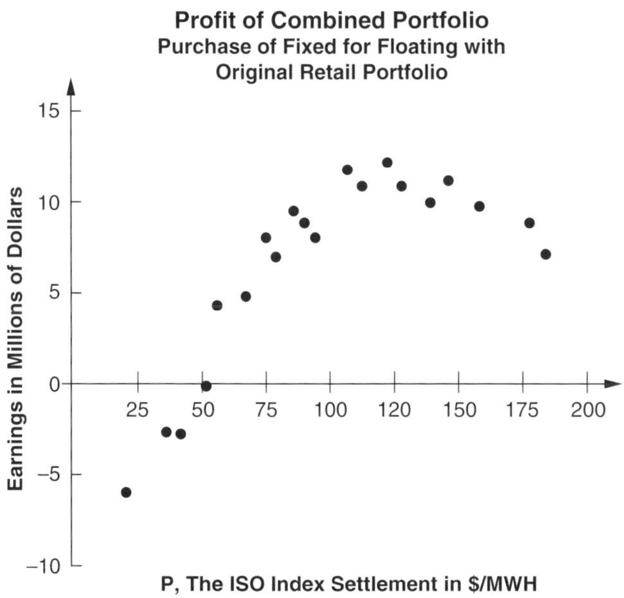

Figure 10.17

approximation to the data, one cannot hedge with the fixed for floating instrument.

- As noted above, the best hedge with the fixed for floating instrument would fully hedge against a hypothetical portfolio with a profit given by the best linear approximation. The deviations from a constant profit of the hedged portfolio are the same as the deviations of the unhedged portfolio from the best linear approximation. If those deviations are large, then hedging will not be effective.
- Also note that in the case that the deviations from the best linear approximation are large, Corr(R,B) is close to 0. Recall that the correlation gives a measurement of how well data fit to a line and if they do not fit well, then the correlation is close to 0.
- For those familiar with regression, the hedge is explained by the regression formula. Regressing the retailer portfolio against the baseload ISO index provides the following:

$$R = aB + Z$$

where a is the regression coefficient and Z is the error term. Adding a baseload instrument provides the following:

$$R + eB = aB + Z + eB = (a + e)B + Z$$

One cannot hedge against the error, Z, since it is completely uncorrelated with the variable B. However, it is possible to hedge against the component of R that is explained by the baseload variable. The best hedge eliminates the uncertainty; set $e^* = -a$. The variance of the hedged portfolio is identically the variance of Z. Low variance in Z is equivalent to a correlation between R and B that is close to 1 or negative 1.

- A further generalization of the geometric interpretation of the optimal hedge is the following. The image of the optimal hedging portfolio should have a shape that is defined by reflecting the best curve fit to the outcomes about a horizontal axis on the graph. If there is not a good curve that fits the data or if the best fit is very different from a line, instruments with nonlinear payoffs are required to develop a good hedge.

10.4.3 Use of Options

A subject that receives a lot of attention in risk management literature is the assessment and use of options. We provide a qualitative description of the products and indicate their use in the retailer portfolio as depicted in Figure 10.15.

Figures 10.18 and 10.19 provide payoffs for holders of calls and puts.

From the shape of the payoff functions it is apparent that these instruments are better suited for mitigating the risk of the retailer's portfolio than the fixed for

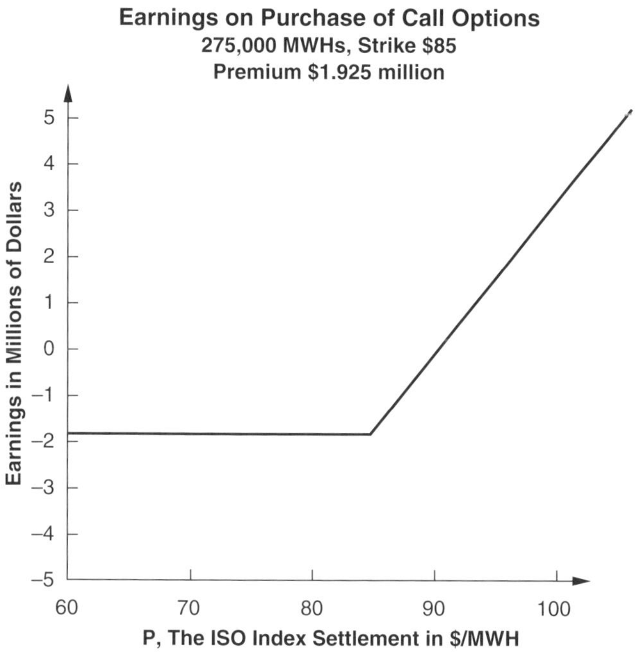

Figure 10.18

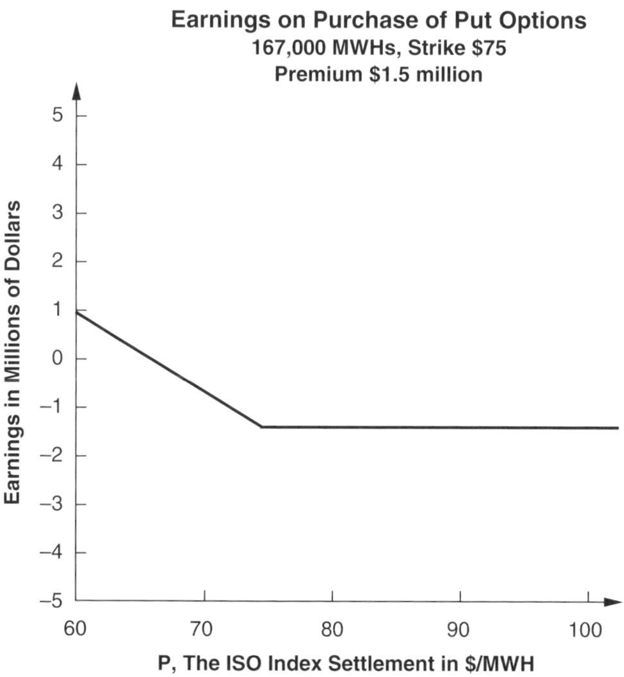

Figure 10.19

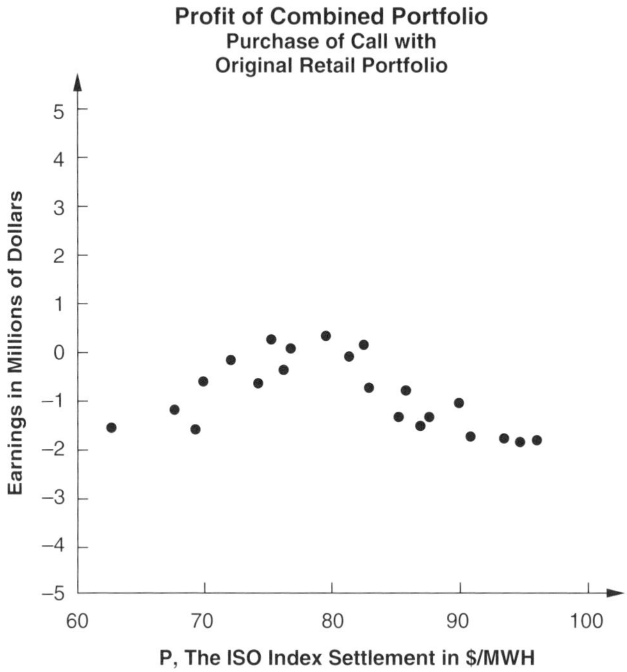

Figure 10.20

floating instruments. The slope of the call option is positive in the region where the slope of the retailer's portfolio earnings is negative, and vice versa for the put. Indeed, we could put these two instruments together and form an instrument that has a very good shape for mitigating the risk of the above-described portfolio.

The graph in Figure 10.20 is the payoff of the retailer portfolio in Figure 10.15 along with the call option displayed in Figure 10.18. One can see from the graph that the instrument does indeed mitigate uncertainty. The range of outcomes has been considerably narrowed to between −2 million and 400 thousand dollars. Unfortunately, since call options are expensive, the cost of mitigating the risk is too great. The associated erosion of profits leaves a very undesirable portfolio.

10.4.4 Summary

There are many more cases and issues to consider. The additional risk in the capacity payment could be presented. Portfolios with customers that pay different prices and have different contract structures could be considered. Quantification of risks could be addressed. All of these topics are of interest; a full treatment is beyond the scope of this text. The essential difficulties that retailers confront in managing their risks have been addressed. The concluding remarks below summarize the presentation and discuss the open issues.

- The central problem for the retailer is the inability to manage risk associated with uncertainty in customer base. The retailer is unable to purchase a fixed for floating contract because sensitivity to the scenario in which falling prices cause customers to depart and the value of the fixed for floating instrument falls sharply. For this reason it is not prudent for the retailer to back up its expected load commitment with market instruments. While this section comes to this conclusion for the retailer's energy needs, the same analysis would yield the same conclusion for the retailer's capacity needs. Accordingly, portfolio positions of retailers are often far out of balance. To some extent, this explains the illiquidity of power markets beyond short time horizons; retailers are unwilling to participate with such uncertainty in their customer base.
- Aside from uncertainty in customer base, retailers face a wide range of risks that cannot be mitigated. These include day-ahead ISO index risk, intraday pricing risk, risk due to imbalance penalties, capacity penalty risk, capacity price risk, and volumetric demand risks for both network costs and revenues. The volumetric demand risks are due to errors in load forecasting for a known customer base and are separate from those associated with the uncertainty in customer base. The analysis of this chapter only addresses the day-ahead ISO index risk because there are no market instruments available for addressing the other risks. Even for the day-ahead ISO index risk with customer certainty, the instruments have limitations. The difference between the load profiles of the customer base and the load profiles of the market instruments limits their use as hedging instruments. Profiles of the market instruments are sold in blocks, whereas load profiles of the customer base vary considerably across hours.
- To minimize intraday price adjustment risks as well as imbalance risks, retailers must develop the same load forecasting skills as utilities in a traditional market environment. In addition to these skills retailers must also develop econometric capabilities that address several issues including pricing, customer sensitivity to pricing, and portfolio risk analysis.
- As pointed out earlier in this chapter, the type of contracts that the retailer signs with its customer base structures the risk in its portfolio. The least risky contract is the contract that is indexed to the day-ahead ISO price. These contracts carry no day-ahead price or load risk. Intraday changes, imbalance penalties and capacity risk, and volumetric demand risks still are present. Unfortunately, there are few customers willing to sign such contracts; customer preference heavily leans toward fixed-price contracts.
- A subject that receives a lot of attention in the risk management literature is the use of options in hedging risk. There are two types of options available, monthly options and daily options. The monthly options may provide a degree of risk mitigation for a changing customer base. Daily options provide a degree of risk mitigation for the daily customer demand fluctuations during delivery month. Neither of these instruments allows for a complete mitigation of risk. They have the same drawback as the fixed for floating contracts: the

inability to profile the financial instrument with the customer load profile. In practice, purchasing options is often too costly for a retailer.

- The equations for risk in this chapter provide qualitative information on the risk components and how they impact one another. However, there are no quantitative solutions. Providing quantitative solutions requires specifying the joint probability distributions for each random variable and explicitly solving for the standard deviations. Joint probabilities are probability functions across several random variables that describe the interaction of the random variables with each other. While there are standard techniques for setting joint probabilities, there are no absolute laws as one finds in physics. Quantitative analysis requires a good deal of mathematical sophistication, and yet setting models and probability distributions requires incorporating business judgment that is not at all quantitative. Aside from setting joint probability distributions, equations must be solved. The equations are more complicated than those presented in this chapter. They include a basket of customer positions across several delivery months as well as a more complex set of market instruments than has been analyzed here. The resulting equations must be solved with the use of computer simulations. Quantitative analysts must be familiar with such methods of solution. In practice, there are limitations to quantitative analytics because of unavoidable errors associated with modeling assumptions and inadequate data.

10.5 RISK MANAGEMENT IN MIDTERM MARKETS: POWER PRODUCERS

In this section we proceed through the analysis of the previous section from the perspective of power producers. There are similarities as well as differences between the risks of producers and retailers. The similarities are in the risks that each market participant confronts and the limitations of the market instruments in addressing these risks. This section first parallels the section on retailer risk to identify the similarities. Later, differences are presented and discussed. Differences arise because of a producer's ability to address structural volumetric risk through maintenance and ownership of many units.

10.5.1 Description of a Producer's Portfolio

We begin with the necessary notation. The future value of a single generation unit over a single delivery month is the following:

$$W = (P - H(G + g) - V)M + D - I + A + K - (S + z)$$

All capitalized letters are random variables, whereas small letters are constant. The variables are defined as follows.

- M is the number of MWH that the producer sells over the delivery month.
- P is the weighted ISO index. The weighting is similar to the retailer case, but MWH sold is used instead of load.
- H is the averaged heat rate, averaged over all MWH sold.
- G is the weighted fuel price. As with power, gas is indexed to a delivery point, except that there is a daily price as opposed to hourly price. The weighting is with respect to daily consumption and is applied to the daily fuel index.
- V is the variable operations and maintenance requirement.
- D represents revenues from intraday schedule adjustments.
- I represents imbalance penalties.
- A is ancillary service revenue over the delivery month. This includes any penalties for nondelivery of promised ancillary services.
- S is the uncertain charges in operating and maintaining the generation units and includes the cost of unscheduled outages.
- K represents the capacity revenues, which is the net of capacity payments received in the capacity markets and penalties for unavailability of capacity that has been sold.
- g is the additional charge of delivering fuel to the unit through a fuel transportation system.
- z is the fixed charge of maintaining the generation units and includes scheduled maintenance charges as well as debt servicing.

The first term of the expression is the earnings from energy sales in the day-ahead ISO market. The term $(P - H(G + g) - V)$ is the difference between the ISO index and production costs given in \$/MWH. The second term provides the difference between energy revenues in intraday markets and imbalance penalties. The third term is the future revenues from selling ancillary services to the ISO. Finally, the fourth term is the cost of operating and maintaining the capacity generation along with the benefit of selling the capacity into the capacity market.

The general case of a producer with several generation units is:

$$W = \sum_{i=1}^{m} W_i$$

The variance of W becomes:

$$\text{Var}(W) = \sum_{i=1}^{m} \text{Var}(W_i) + 2\sum_{j=1}^{m-1} \sum_{k=j+1}^{m} \text{Cov}(W_j, W_k)$$

EXAMPLE 1. *A Single Unit with Fixed Fuel Costs and Fixed Output*

In this example we apply the above steps to the portfolio assuming fixed load and output. Following Example 1 of Section 10.4, this example demonstrates that the market instruments are insufficient for mitigating risk because the profile of the unit's output does not match the profile of the market instruments.

Step 1. Develop a set of simplifying assumptions.

The first set of assumptions is the following.

- There is a single unit.
- The only variable is P. All other terms are constant.

With these assumptions we consider only the first term of the future value.

$$W = (P - H(G + g) - V)M$$

$$\text{Var}(W) = M^2\text{Var}(P)$$

Step 2. Determine the set of market instruments that are incorporated in the portfolio

We select a baseload instrument. In this case, the producers sells the baseload for fixed.

$$W + e(f - B) = (P - H(G + g) - V)M + e(f - B)$$

Step 3. Write down the expression for the variance of the portfolio with market instruments included and develop further simplifying assumptions.

Using Formulas 1, 2, and 3 from Section 10.3, the variance is the following:

$$\begin{aligned}\text{Var}(W + e(f - B)) &= \text{Var}(W) + e^2\text{Var}(f - B) + 2e\text{Cov}(W, f - B)\\ &= \text{Var}(W) + e^2\text{Var}(B) - 2e\text{Cov}(W,B)\\ &= M^2\text{Var}(P) + e^2\text{Var}(B) - 2e\text{Cov}((P - H(G + g) - V)M, B)\\ &= \text{Var}(B)e^2 - 2M\text{Cov}(P,B)e + M^2\text{Var}(P)\end{aligned}$$

The last equality emphasizes that the result is a quadratic function in e. The variance is a parabola with respect to the value e. One could mitigate the price risk by setting the expression to 0. Using the quadratic formula, we solve for e.

$$\begin{aligned}e &= \frac{M\text{Cov}(P,B) \pm \sqrt{(M\text{Cov}(P,B))^2 - M^2\text{Var}(P)\text{Var}(B)}}{\text{Var}(B)}\\ &= \frac{M\text{Cov}(P,B) \pm \sqrt{(M\text{Corr}(P,B))^2\text{Var}(P)\text{Var}(B) - M^2\text{Var}(P)\text{Var}(B)}}{\text{Var}(B)}\\ &= \frac{M\text{Cov}(P,B) \pm \sqrt{(\text{Corr}(P,B)^2 - 1)M^2\text{Var}(P)\text{Var}(B)}}{\text{Var}(B)}\\ &= \frac{M\text{Cov}(P,B) \pm M\text{SD}(P)\text{SD}(B)\sqrt{(\text{Corr}(P,B)^2 - 1)}}{\text{Var}(B)}\end{aligned}$$

The only time that the term inside the square root is not negative occurs when $\text{Corr}(P,B) = 1$. In this case B and P are identical, $B = P$, and the above equation reduces to $e = M$. The retailer would sell M MWH of baseload; producers would be able to eliminate their day-ahead energy cost risk.

For the general case, B and P differ because the load profile of the unit output differs from the baseload profile of the market instrument. The situation is similar to that of the retailer.

Step 4. Solve for the quantity of the market instruments that minimizes the risk.

The value of e that minimizes the variance is the value that gives the vertex of the graph of the variance. In general the vertex of the equation

$$\text{Variance}(e) = A_2e^2 + 2A_1e + A_0$$

is given at the point

$$e^* = -\frac{A_1}{A_2}$$

In the above case, $A_1 = -\text{Cov}(P,B)M$ while $A_2 = \text{Var}(B)$.

This gives the following optimal value, e^*.

$$\begin{aligned} e^* &= \frac{\text{Cov}(P,B)L}{\text{Var}(B)} \\ &= \text{Corr}(P,B)L\frac{\text{SD}(P)}{\text{SD}(B)} \end{aligned}$$

Step 5. Determine the value of the risk associated with the optimal hedge.

$$\begin{aligned} \text{SD}(R + e^*(B - f)) &= \sqrt{\text{Var}(B)e^2 - 2\text{Cov}(P,B)Me + M^2\text{Var}(P)} \\ &= \sqrt{\frac{\text{Cov}(P,B)^2M^2 - 2\text{Cov}(P,B)^2M^2}{\text{Var}(B)} + M^2(\text{Var}(P))} \\ &= \sqrt{\frac{M^2\text{Var}(P) - \text{Cov}(P,B)^2M^2}{\text{Var}(B)}} \\ &= M\sqrt{\text{Var}(P) - \text{Corr}(P,B)^2\text{Var}(P)} \\ &= \sqrt{1 - \text{Corr}(P,B)^2}\,M\text{SD}(P) \end{aligned}$$

The result is identical to that of the retailer in which M, the unit output, assumes the same role as L, the retailer load. The remarks and commentary from Example 1 in Section 10.4 apply to this example as well.

EXAMPLE 2. *A Single Unit with Uncertain Output*

This example applies the approach of the preceding example to the case in which output is uncertain and all other parameters are fixed. We show that the fixed for floating market instrument is unable to hedge any risk in this portfolio. This follows Example 2 of Section 10.4.

Step 1. Develop a set of simplifying assumptions.

- The only uncertain variable is M, the unit output.
- $P = B = f$. The index is the same as the baseload contract. Both are fixed at the forward market value.

Step 2. Determine the set of market instruments that are incorporated in the portfolio.

The only market contract considered is the baseload fixed for floating. Neglecting all fixed terms, the future value becomes the following:

$$W + e(f - B) = (P - H(G + g) - V)M + e(f - B)$$

Step 3. Write down the expression for the variance of the portfolio with market instruments included and develop further simplifying assumptions.

Calculating the variance of W with Formulas 1, 2, and 3 from Section 10.3 gives the following:

$$\text{Var}(W) = ((P - H(G + g) - V)^2\text{Var}(M)$$

As before, we examine a portfolio with the addition of fixed for floating contracts and try to determine the optimal hedging volume, e.

$$\text{Var}(W + e(f - B)) = \text{Var}(W) + e^2\text{Var}(f - B) + 2e\text{Cov}(W, f - B) = \text{Var}(W)$$

since the only random variable is W, all other terms vanish.

Step 4. Solve for the quantity of the market instruments that minimize the risk.

The portfolio risk is independent of the choice of e; the value remains the same no matter what value of e is selected.

Step 5. Determine the value of the risk associated with the optimal hedge.

$$\text{SD}(W + e(f - B)) = \text{SD}(W) = (P - H(G + g) - V)\text{SD}(M)$$

The risk in this portfolio depends on the standard deviation of the output, and one is unable to mitigate this risk with a fixed for floating instrument.

EXAMPLE 3. *A Single Unit with Output Uncertainty and Price Uncertainty*

This example follows Example 3 in Section 10.4.

Step 1. Develop a set of simplifying assumptions.

- $W = (P - H(G + g) - V)M$
- The only variables are P and M. All other parameters are fixed.

Step 2. Determine the set of market instruments that are incorporated in the portfolio.

The only market contract considered is the baseload fixed for floating contract. Neglecting all fixed terms, the future value becomes the following:

$$W + e(f - B) = (P - H(G + g) - V)M + e(f - B)$$

Step 3. Write down the expression for the variance of the portfolio with market instruments included and develop further simplifying assumptions.

With Formulas 1, 2, and 3 from Section 10.3, the variance is the following:

$$\text{Var}(W) = (H(G + g) - V)^2\text{Var}(M) + \text{Var}(MP) - 2(H(G + g) - V)\text{Cov}(M,MP)$$

As before, we examine a portfolio with the addition of fixed for floating contracts and try to determine the optimal hedging volume, e.

$$\begin{aligned}\text{Var}(W + e(f - B)) &= \text{Var}(W) + e^2\text{Var}(f - B) + 2e\text{Cov}(W,f - B)\\ &= \text{Var}(W) + e^2\text{Var}(B) - 2e\text{Cov}(W,B)\\ &= \text{Var}(B)e^2 - 2\text{Cov}(W,B)e + \text{Var}(W)\end{aligned}$$

Next, we divert momentarily from the steps for selecting e^* and determine whether it is possible to eliminate risk. Elimination of risk requires setting the above expression to 0. As before, we note that the variance is quadratic in e and use the quadratic formula to set the variance to 0 by solving for e.

$$
\begin{aligned}
e &= \frac{\mathrm{Cov}(W, B) \pm \sqrt{\mathrm{Cov}(W,B)^2 - \mathrm{Var}(W)\mathrm{Var}(B)}}{\mathrm{Var}(B)} \\
&= \frac{-\mathrm{Cov}(W, B) \pm \sqrt{\mathrm{Corr}(W,B)^2\mathrm{Var}(W)\mathrm{Var}(B) - \mathrm{Var}(W)\mathrm{Var}(B)}}{\mathrm{Var}(P)} \\
&= \frac{\left(\mathrm{Corr}(W, B) \pm \sqrt{\mathrm{Corr}(W, B)^2 - 1}\right)\mathrm{SD}(W)\mathrm{SD}(B)}{\mathrm{Var}(P)} \\
&= \left(\mathrm{Corr}(W, B) \pm \sqrt{\mathrm{Corr}(W,B)^2 - 1}\right)\frac{\mathrm{SD}(W)}{\mathrm{SD}(B)}
\end{aligned}
$$

The term in the square root is non-negative only if the cash flow, W, and the baseload ISO index, B, are fully correlated or fully uncorrelated. Under all other circumstances, it is not possible to completely mitigate the risk.

Step 4. Solve for the quantity of the market instruments that minimize the risk.

As before the value of e that provides the minimum risk is found by determining the value of e where the vertex of the parabola gives the following:

$$
\begin{aligned}
e^* &= \frac{\mathrm{Cov}(B,W)}{\mathrm{Var}(B)} \\
&= \mathrm{Corr}(B,W)\frac{\mathrm{SD}(W)}{\mathrm{SD}(B)}
\end{aligned}
$$

Step 5. Determine the value of the risk associated with the optimal hedge.

The risk associated with this choice for e is the following:

$$
\begin{aligned}
\mathrm{SD}((W + e^*(B - f)) &= \sqrt{\mathrm{Var}(B)e^{*2} - 2\mathrm{Cov}(W,B)e^* + \mathrm{Var}(W)} \\
&= \sqrt{\frac{\mathrm{Var}(B)\mathrm{Cov}(B,W)}{\mathrm{Var}(B)^2} - \frac{2\mathrm{Cov}(W,B)^2}{\mathrm{Var}(B)} + \mathrm{Var}(W)} \\
&= \sqrt{\mathrm{Var}(W) - \frac{\mathrm{Cov}(W,B)^2}{\mathrm{Var}(B)}} \\
&= \sqrt{\mathrm{Var}(W) - \mathrm{Corr}(W,B)^2\mathrm{Var}(W)} \\
&= \sqrt{1 - \mathrm{Corr}(W,B)^2}\,\mathrm{SD}(W)
\end{aligned}
$$

Similar to the retailer's case, the result depends on the correlation between the future earnings of the portfolio and the value of the market instrument. In the preceding section we saw that the correlation between the value of the retailer's portfolio and the value of the market instrument was dependent on customer response to market changes. The corresponding correlation for the producer differs. We explore this below.

10.5.2 Further Discussion of Corr(*W*,*B*), Output-Price Uncertainty, and Hedging

This subsection corresponds to the subsection following Example 3 of Section 10.4. In that subsection the correlation, Corr(R,P), is investigated with demand and profit outcomes. A similar approach is taken in the investigation of Corr(W,B), viewing supply and profit outcomes. The approach looks at three cases: a baseload unit, an intermediate unit, and a peaking unit. The analysis of the baseload unit is developed through formulas, while the intermediate and peaking unit cases are depicted through scatter plots. Our assumptions are the following.

- $W = (P - H(G + g) - V)M$
- The hedging instrument is the baseload fixed for floating and the price, f, is \$65/MWH
- There are no forced outages.

EXAMPLE 4. *Baseload*

Recall that volumetric considerations for the retailer are captured through the demand function. The corresponding concept for the producer is the supply function. In this example, we present supply and profit functions for the baseload unit under the following additional assumptions.

- The unit is a coal-fired plant. The total variable cost, fuel and VOM, is not a random variable and is set as follows:

$$H(G + g) - V = \$34/\text{MWH}$$

- Prices do not fall below the total variable cost, so that $P = B$.
- The unit capacity is 500 MW
- The time horizon is a month with 720 hours

With these assumptions the production, M, is a constant.

$$\text{Supply} = M = 360{,}000\,\text{MWH}$$

This equation gives the supply function.

The profit function then becomes the following:

$$\text{Profit} = W = 360{,}000(B - 34)$$

Since W is linear in B, the correlation $\text{Corr}(W,B) = 1$. This portfolio can be completely hedged by selling the fixed for floating instrument. Note that the slope of the profit function is positive, so e^* is negative. Indeed, the optimal value for e^* is −360,000 MWH. The profit on the hedged portfolio becomes the following:

$$\text{Profit with hedge} = W - e^*(B - f) = 360{,}000(65 - 34) = 12.24 \text{ million dollars}$$

Figure 10.21 provides a plot of the profit function for the hedged and unhedged portfolios. The sale of the fixed for floating contract is also shown.

EXAMPLE 5. *An Intermediate Unit*

We assume a CCGT with the following characteristics.

- A heat rate $H = 7.5$ MMBTU/MWH
- VOM = 5.00 \$/MWH
- A capacity of 500 MW

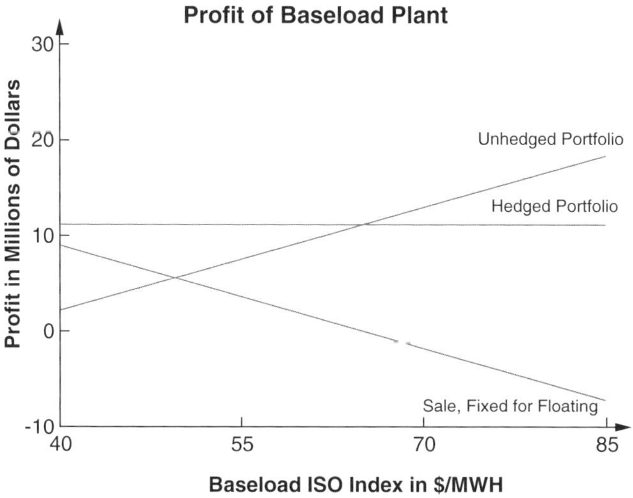

Figure 10.21

The supply relation with the ISO index is depicted through the scatter plot of Figure 10.22.

Figure 10.23 provides a plot of the unhedged earnings along with the best linear approximation to the data. The slope of the approximation is 200,000 MWH/$, indicating that e^* is – 200,000 MWH; a 200,000-MWH sale of the fixed for floating contract. The portfolio with the optimal hedge is seen in the graph of Figure 10.24.

EXAMPLE 6. *A Peaking Unit*

We assume a CT with the following characteristics.

- A heat rate $H = 10.5$ MMBTU/MWH
- VOM = 3.00 $/MWH
- A capacity of 200 MW

The supply relation with the ISO index is depicted through the scatter plot of Figure 10.25.

Figure 10.26 provides a plot of the unhedged earnings along with the best linear approximation to the data. The slope of the approximation is 40,000 MWH/$, indicating that e^* is –40,000 MWH; a 40,000 MWH sale of the fixed for floating contract. The portfolio with the optimal hedge is seen in the graph of Figure 10.27.

The examples demonstrate some properties that we remark upon.

Remarks

- The baseload unit has perfect correlation with the market instrument, and the hedge works well.
- The hedge for the CT is not very effective, as the uncertainty of profit in the hedged CT is nearly the same as the uncertainty in the unhedged CT. The best

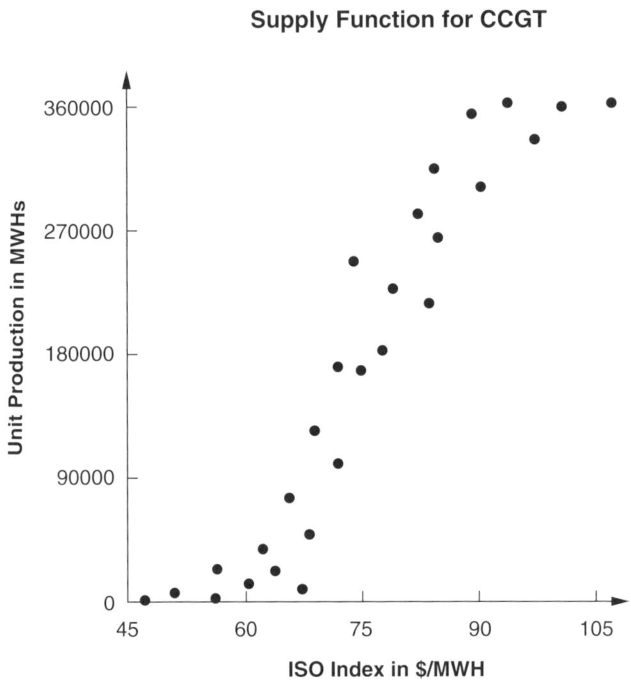

Figure 10.22

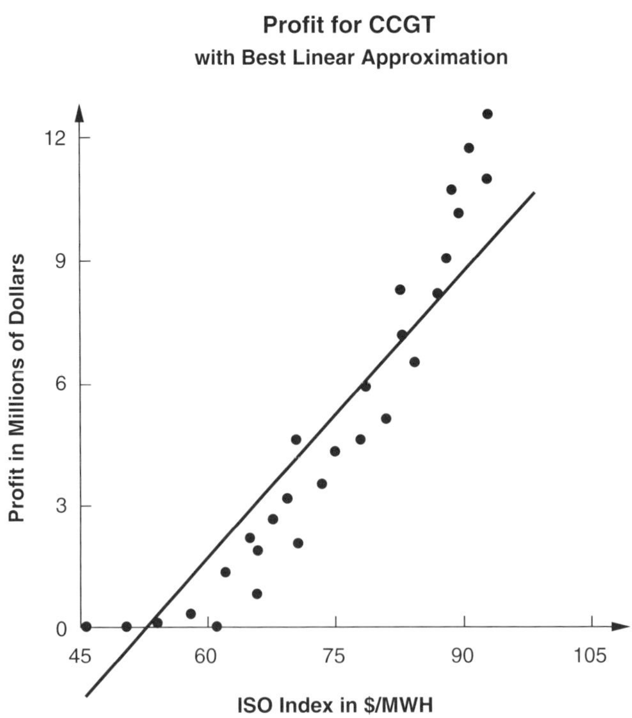

Figure 10.23

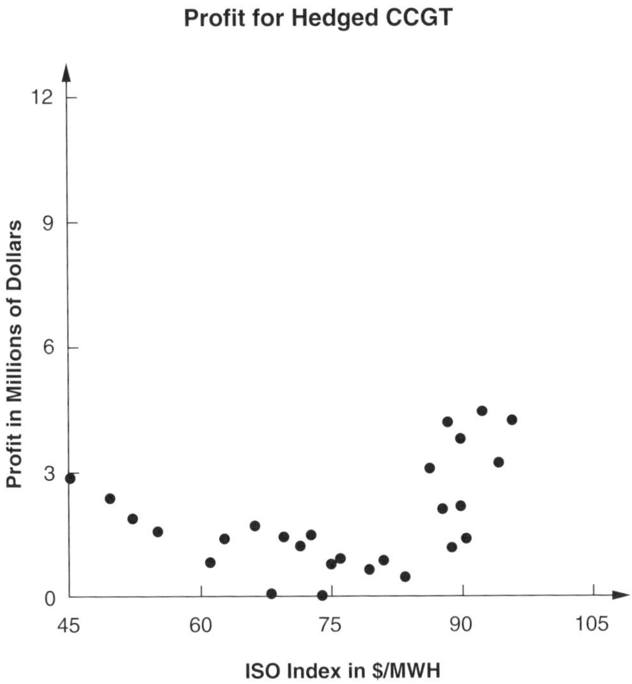

Figure 10.24

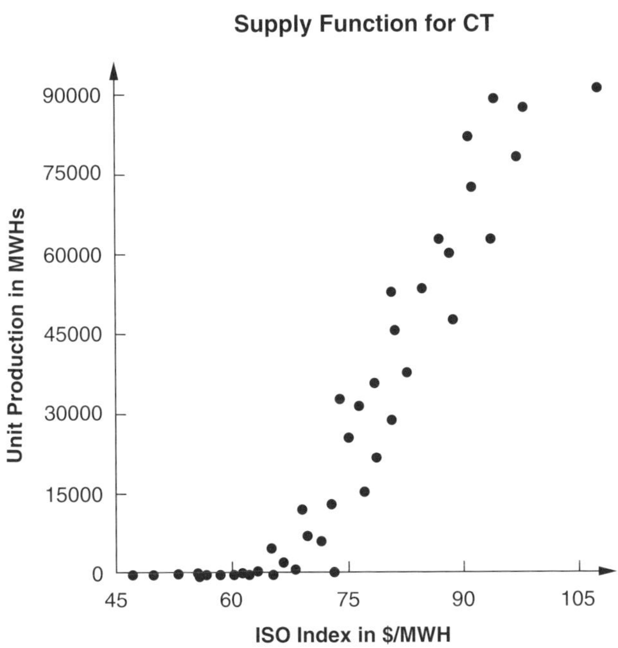

Figure 10.25

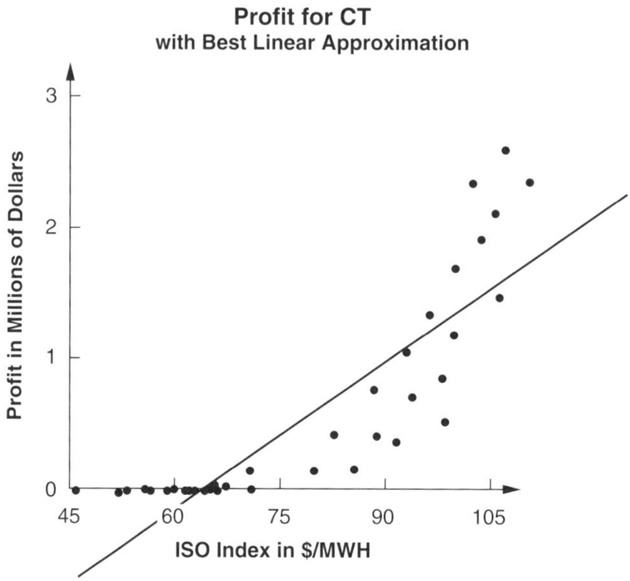

Figure 10.26

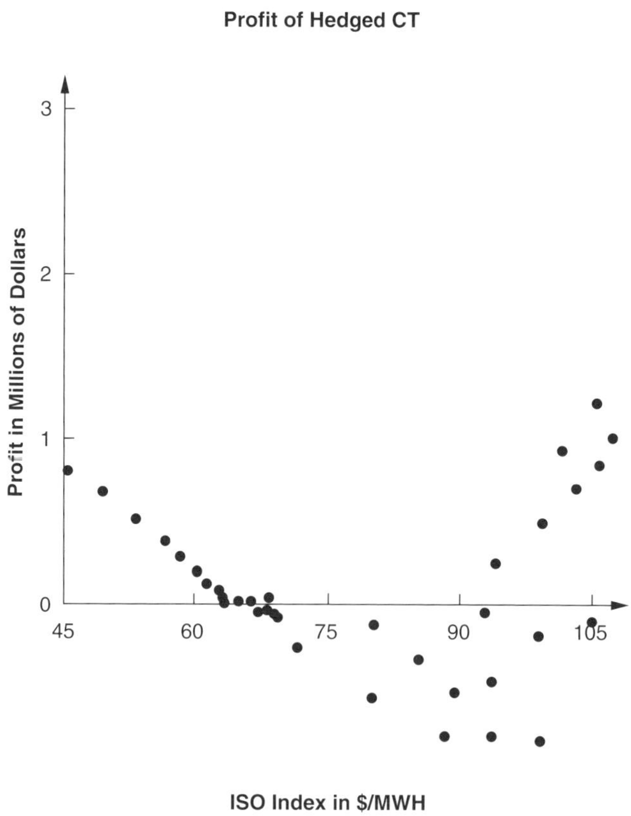

Figure 10.27

linear approximation for the CT is not a very good fit. While this may be partially attributable to our use of a baseload instrument as opposed to an on-peak instrument, this is also the case for on-peak instruments. CTs are difficult to hedge with fixed for floating instruments. The correlation between the monthly ISO index and the profit of the CT is normally quite low.

- The effectiveness of the hedge for the CCGT falls in between the cases for the baseload unit and the CT.
- Options are much more effective hedging instruments for CTs and CCGTs than fixed for floating contracts. A good choice among power options is the sale of a call option since the earnings on the sale of a call are complementary to the shape of the profit relations for both CTs and CCGTs. We invite the reader to generate a hedge and profit relation.

10.5.3 Forced Outages

The reader may have noted a critical element that is missing from the above analysis, forced outages. This aspect is next introduced into the analysis. Because of forced outages the producer may not realize all the potential profits that are achievable in the market. The effect of forced outages on the earnings plots is that some of the profit outcomes would be reduced. This is most easily seen for the baseload case. Instead of having all profits aligned along the line as in the graph in Figure 10.21, some outcomes would fall below the line, as depicted in Figure 10.28. There would

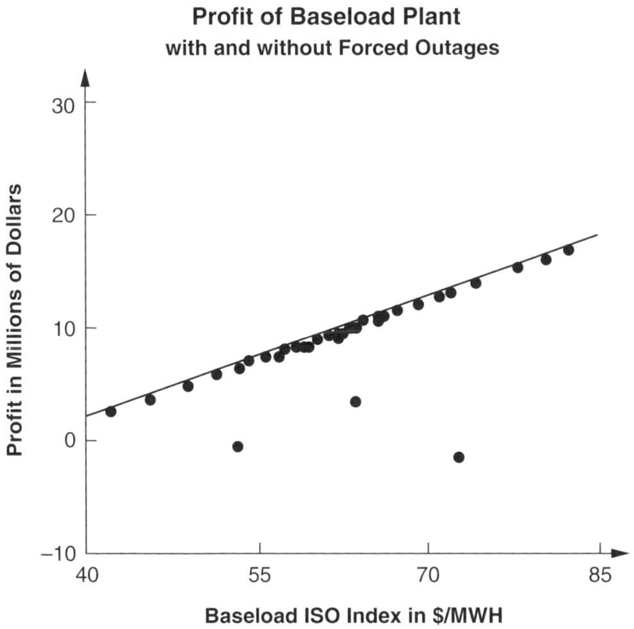

Figure 10.28

no longer be perfect correlation between the ISO index and the profit outcomes, resulting in less effective hedging.

The decrease in the correlation, Corr(W,B), due to forced outages occurs in other units just as demonstrated in the above case with a baseload unit. The decrease in correlation is minimal for units with low forced outage rates and more significant as forced outages rates increase.

It is our goal to quantify the affect of the forced outage on the correlation. Some terminology is required. The availability of a unit over a month's time frame is the fraction of the time that the unit is operable. An availability of 1 means that the unit is operable over all hours of the month, even though it may not be dispatched. The availability is a random variable.

Assume that the availability is independent of both the baseload ISO index, B, and the future value of the portfolio, W. Then the correlation Corr(W,B) is given by the following formula:

$$\mathrm{Corr}(W,B) = \mathrm{E(Av)}\frac{\mathrm{SD}(W^1)}{\mathrm{SD}(W)}\mathrm{Corr}(W^1,B)$$

where

- E(Av) is the expected availability, $0 < \mathrm{E(Av)} < 1$
- W^1 is the future earnings of the portfolio under the assumption that the unit is always available.

In the next subsection we further develop the standard deviation of the future cash flow, SD(W), and refine the correlation. There it is seen that the expression $\mathrm{E(Av)}\frac{\mathrm{SD}(W^1)}{\mathrm{SD}(W)}$ is less than 1 and risk mitigation is more effective when the expression is close to 1. This illustrates an important point. Operations impact the risk of a producer's portfolio, and investing in maintenance to minimize forced outages is a critical element of a risk mitigation strategy. This concept is further developed below.

10.5.4 Further Discussion of SD(W)

Following Example 3, aside from Corr(W,B) another factor in the risk of the hedged portfolio is SD(W), the standard deviation of the future cash flow. In this section the standard deviation is further analyzed. The assumptions are the following.

- There is a single unit.
- $W = (P - H(G + g) - V)M$
- The only variables are P and M. All other parameters are fixed.
- $M = M^1\mathrm{Av}$
- Av is a random variable and represents the availability. Av is independent of all other variables.
- M^1 is the production assuming full availability.

Recall Formula 4 at the end of Section 10.3.

$$\text{Var}(XY) = \text{Var}(X)\text{Var}(Y) + [E(X)]^2\text{Var}(Y) + [E(Y)]^2\text{Var}(X)$$

whenever X and Y are two independent random variables.

Setting $X = (P - H(G + g) - V)M^1$, $Y = \text{Av}$ and applying the above formula gives the following result for Var(W):

$$\begin{aligned}\text{Var}(W) = {} & \text{Var}[(P - H(G + g) - V)M^1]\text{Var}(\text{Av}) \\ & + [E(P - H(G + g) - V)M^1)]^2\text{Var}(\text{Av}) \\ & + [E(\text{Av})]^2\text{Var}[(P - H(G + g) - V)M^1]\end{aligned}$$

Taking the square root gives the standard deviation, $\text{SD}(W) = \sqrt{\text{Var}(W)}$.

Variance in Av adds substantial risk to the portfolio. Referring to the final results of Example 3, this impacts the effectiveness of hedging. In the following subsection we demonstrate how to reduce the variance of Av through diversification.

Remarks

- Minimizing the variance of the output is part of an overall mitigation strategy.
- Typically, both the earnings from sales of power and the variance of these earnings are low for peaking units. Accordingly, the above standard deviation is low for a peaking unit. Peaking units must earn value from capacity payments and by selling into the ancillary service markets. However, similar to a baseload unit's risk in energy earnings, there is risk to the income from capacity revenues and ancillary service revenues due to unavailability.
- Returning to the correlation, Corr(W,B), from the end of the previous subsection, one finds that the expression

$$\text{E}(\text{Av})\frac{\text{SD}(W^1)}{\text{SD}(W)}$$

  is close to 1 whenever the variance of Av is small. Conversely, this expression is close to 0 whenever the variance of Av is large. This demonstrates that risk mitigation by market instruments is most effective whenever the availability has little variance.

10.5.5 Diversification with Multiple Units

The previous subsection demonstrates that decreasing the uncertainty in output also decreases the risk in a portfolio. In this subsection, we demonstrate that operating a fleet of units decreases the uncertainty in output through diversification.

Indeed, diversification is a common practice in risk management. The concept is accepted as common wisdom that is embodied in the expression, "Don't put your eggs in one basket." While the common wisdom came about through experience, there is a formalized mathematical principle at work. We share this principle and apply it to risk management in midterm markets.

A property of the standard deviation of a sum of random variables is the following:

$$\mathrm{SD}(X+Y) \le \mathrm{SD}(X) + \mathrm{SD}(Y)$$

For multiple variables the property is the following:

$$\mathrm{SD}\left(\sum_{j=1}^{n} X_j\right) \le \sum_{j=1}^{n} \mathrm{SD}(X_j)$$

In the case when the random variables X_j are independent, the following formula applies:

$$\mathrm{SD}\left(\sum_{j=1}^{n} X_j\right) = \sqrt{\sum_{j=1}^{n} \mathrm{Var}(X_j)} < \sum_{j=1}^{n} \mathrm{SD}(X_j)$$

In the case when the X_j are independent and have the same variance the above formula becomes the following:

$$\mathrm{SD}\left(\sum_{j=1}^{n} X_j\right) = \sqrt{n}\,\mathrm{SD}(X_k)$$

where k is an arbitrary index between 1 and n.

This is meaningful to the risk manager because it states that the risk of a portfolio is no greater than the sum of the risks of the individual portfolio elements. In the case when portfolio elements behave independently, the uncertainty of the portfolio is strictly less than the summed uncertainty of each element. The difference may be substantial. This happens through diversification because uncertainties offset each other when the portfolio is combined. We apply this concept to a portfolio of multiple units.

EXAMPLE 7. *Multiple Units with Output Uncertainty and Price Uncertainty*

In this example the five-step method for mitigating risk is applied to the case when there are multiple units. The purpose is to determine how the risk profile changes as more units are integrated into the portfolio. The result is an analytic formula that parallels Example 3. The reader wishing to see numerical results can skip to Example 8, where the formulas are applied to a numerical investigation.

Step 1. Develop a set of simplifying assumptions.

- There are n units with the same operating characteristics, including availability.

- The bidding for each unit is identical.
- $W = \sum_{j=1}^{n} W_j$
- $W_j = (P_j - H(G + g) - V)M_j$
- The only variables are P_j and M_j. All other parameters are fixed.
- For every j, $M_j = M_j^1 \mathrm{Av}_j$
- M_j^1 is the output assuming full availability.
- Av_j is the availability of unit j and is independent of all other random variables including the availability factor of other units. This assumption means that one unit's availability has no impact on another unit's availability. Because of the assumption that operating characteristics of the units are identical, the expected value and standard deviation are the same for all of the Av_j.

Because of the assumption that operating characteristics of the units are identical, the expression $(P_j - H(G + g) - V)M_j^1$ is identical for each j. With these assumptions, W is expressed as follows:

$$\begin{aligned} W &= \sum_{j=1}^{n} W_j \\ &= \sum_{j=1}^{n} (P_j - H(\mathrm{G} + g) - V)M_j^1 \mathrm{Av}_j \\ &= (P_k - H(G + g) - V)M_k^1 \sum_{j=1}^{n} \mathrm{Av}_j \end{aligned}$$

where k represents an arbitrary unit.

For convenience we set $\mathrm{Av} = \sum_{j=1}^{n} \mathrm{Av}_j$. Note that $\mathrm{E}(\mathrm{Av}) = n\mathrm{E}(\mathrm{Av}_k)$.

Step 2. Determine the set of market instruments that are incorporated in the portfolio.

The only market contract considered is the baseload fixed for floating. Neglecting all fixed terms, the future value becomes the following.

$$W + e(f - B) = (P_k - H(G + g) - V)M_k^1 \mathrm{Av} + e(f - B)$$

Step 3. Write down the expression for the variance of the portfolio with market instruments included and develop further simplifying assumptions.

As before, we examine portfolio with the addition of fixed for floating contracts and try to determine the optimal hedging volume, e.

$$\begin{aligned} \mathrm{Var}(W + e(f - B)) &= \mathrm{Var}(W) + e^2\mathrm{Var}(f - B) + 2e\mathrm{Cov}(W, f - B) \\ &= \mathrm{Var}(W) + e^2\mathrm{Var}(B) - 2e\mathrm{Cov}(W, B) \\ &= \mathrm{Var}(B)e^2 - 2\mathrm{Cov}(W, B)e + \mathrm{Var}(W) \end{aligned}$$

As with Example 3, it is impossible to mitigate risk unless $\mathrm{Corr}(W, B) = 1$.

Step 4. Solve for the quantity of the market instruments that minimize the risk.

As before, the value of e that provides the minimum risk is found by determining the value of e where the vertex of the parabola gives the following:

$$e^* = \frac{\mathrm{Cov}(B,W)}{\mathrm{Var}(B)}$$
$$= \mathrm{Corr}(B,W)\frac{\mathrm{SD}(W)}{\mathrm{SD}(B)}$$

This is the same result as in Step 4 of Example 3.

Step 5. Determine the value of the risk associated with the optimal hedge.

The risk associated with this choice for e is the following:

$$\begin{aligned}\mathrm{SD}(W + e^*(B - f)) &= \sqrt{\mathrm{Var}(B)e^{*2} - 2\mathrm{Cov}(W,B)e^* + \mathrm{Var}(W)}\\ &= \sqrt{\frac{\mathrm{Var}(B)\mathrm{Cov}(B,W)}{\mathrm{Var}(B)^2} - \frac{2\mathrm{Cov}(W,B)^2}{\mathrm{Var}(B)} + \mathrm{Var}(W)}\\ &= \sqrt{\mathrm{Var}(W) - \frac{\mathrm{Cov}(W,B)^2}{\mathrm{Var}(B)}}\\ &= \sqrt{\mathrm{Var}(W) - \mathrm{Corr}(W,B)^2\mathrm{Var}(W)}\\ &= \sqrt{1 - \mathrm{Corr}(W,B)^2}\,\mathrm{SD}(W)\end{aligned}$$

This result is identical to that of Example 3. Using the results of the previous subsections, we have the following.

$$\mathrm{Corr}(W,B) = \mathrm{E}(\mathrm{Av})\frac{\mathrm{SD}(W_k^1)}{\mathrm{SD}(W)}\mathrm{Corr}(W_k^1,B)$$

$$\mathrm{SD}(W) = \sqrt{\mathrm{Var}(\mathrm{Av})\mathrm{Var}(W_k^1) + [E(W_k^1)]^2\mathrm{Var}(\mathrm{Av}) + [E(\mathrm{Av})]^2\mathrm{Var}(W_k^1)}$$

Next note that $\mathrm{Var}(\mathrm{Av}) = n\mathrm{Var}(\mathrm{Av}_k)$. Substituting into the above expression gives the following:

$$\begin{aligned}\mathrm{SD}(W) &= \sqrt{n}\sqrt{\mathrm{Var}(\mathrm{Av}_k)\mathrm{Var}(W_k^1) + [E(W_k^1)]^2\mathrm{Var}(\mathrm{Av}_k) + n[\mathrm{E}(\mathrm{Av}_k)]^2\mathrm{Var}(W_k^1)}\\ &= \sqrt{n}\sqrt{\mathrm{Var}(\mathrm{Av}_k)[\mathrm{Var}(W_k^1) + [E(W_k^1)]^2] + n[\mathrm{E}(\mathrm{Av}_k)]^2 + \mathrm{Var}(W_k^1)}\end{aligned}$$

The final term in the square root becomes the dominant term as the number of units increases.

$\mathrm{SD}(W)$ as a function of n has interesting behavior. As the number of units, n, increases, the first term $\mathrm{Var}(\mathrm{Av}_k)$ $[\mathrm{Var}(W_k^1) + [E(W_k^1)]^2]$ becomes insignificant in comparison with the second term, $n[\mathrm{E}(\mathrm{Av})]^2\mathrm{Var}(W_k^1)$. The standard deviation, $\mathrm{SD}(W)$ asymptotically approaches the value $n\mathrm{E}(\mathrm{Av}_k)\mathrm{SD}(W_k^1) = \mathrm{E}(\mathrm{Av})\ \mathrm{SD}(W_k^1)$.

Substitution of the result for $\mathrm{SD}(W)$ into the expression $\mathrm{E}(\mathrm{Av})\frac{\mathrm{SD}(W_k^1)}{\mathrm{SD}(W)}$ provides an interesting insight. The expression is less than 1 whenever E(Av) is less than 1 and increases as n increases. In fact, the expression approaches 1 asymptotically and the effect of $\mathrm{Var}(\mathrm{Av}_k)$ vanishes asymptotically. This means that as the number of units increases, $\mathrm{Corr}(W,B)$ also increases and hedging becomes more effective.

This example demonstrates that, by increasing the number of units, one can reduce the risk associated with the variance in unit availability. More discussion of this effect follows.

EXAMPLE 8. *Numerical Investigation*

In this example, numerical assumptions are introduced into Example 7 to get a sense of the magnitude of risk reduction as a result of increasing the number of units. We start by listing the assumptions and then calculate the risk of the minimum risk portfolio from Example 7.

The assumptions are as follows.

- All assumptions from Example 7 apply.
- The units are all extremely efficient CCGTs. For simplicity, we assume that they are baseload units.
- The heat rates are $H = 6.8$ in MMBTU per MWH
- The gas price $G = 7.75$ in dollars per MMBTU
- $g = .25$ in dollars per MMBTU
- $V = 5.00$ in dollars per MWH
- The size of each unit is 400 MW, and there are 720 hours in the month. Then $M_k^1 = 288{,}000$ MWH.
- The expected baseload price, P_k, is 77 in dollars per MWH.
- From the above assumptions, the expected cash flow assuming full availability is the following:

$$E(P_k - (H(G + g)) + V)M_k^1 = 5{,}068{,}800 \text{ dollars}$$

- The variance in the price is $\text{Var}(P_k) = 100$: $\text{SD}(P_k) = 10$ dollars/MWH
- $\text{Var}(P_k - (H(G + g)) + V)M_k^1 = \text{Var}(W_k^1) = 8{,}294{,}400{,}000{,}000$
  $\text{SD}(P_k - (H(G + g)) + V)M_k^1 = \text{SD}(W_k^1) = 2{,}880{,}000$ dollars
- For all units, $E(\text{Av}_j) = .92$
- For all units, $\text{Var}(\text{Av}_j) = .0625$ and $\text{SD}(\text{Av}_j) = .25$
- $\text{Corr}(W_k^1, B) = 1$ (this is the same as $P_k = B$)
- $\text{Corr}(W,B) = E(\text{Av})\dfrac{\text{SD}(W^1)}{\text{SD}(W)}\text{Corr}(W^1,B) = \text{E}(\text{Av})\dfrac{\text{SD}(W^1)}{\text{SD}(W)}$

With these assumptions we can assess the risk of Example 4.

$$\text{SD}(W) = \sqrt{n}\sqrt{\text{VarAv}_k \text{Var}W_k^1 + [EW_k^1]^2\,\text{VarAv} + n[\text{EAv}_k]^2\,\text{Var}W_k^1}$$

$$\text{Var}(\text{Av}_k)\text{Var}(W_k^1) = .0625 \times 8{,}294{,}400{,}000{,}000 = 518{,}400{,}000{,}000$$

$$[EW_k^1]^2\text{Var}(\text{Av}) = 25{,}692{,}733{,}440{,}000 \times .0625 = 1{,}605{,}795{,}840{,}000$$

$$[\text{EAv}_k]^2\text{Var}(W_k^1) = 7{,}020{,}380{,}160{,}000$$

Table 10.4 provides the risk of the future earnings, SD(W), as a function of the number of units, n.

There are several things to note.

- For a single unit the risk assuming full availability is 2,880,000 dollars. This compares to 3,024,000 dollars when availability is considered. The uncertainty of availability adds 144,000 dollars in risk. This represents 4.8 percent of the total risk.
- As the number of units increases, the total risk does not increase linearly. Indeed, the risk of 10 units is substantially less than 10 times the risk of 1 unit.
- The risk of increasing the number of units in the portfolio by 1 decreases as the number of units gets larger. Indeed, by adding 1 more unit from 9 to 10, one increases the risk by 2,649,926 dollars. This is less than the risk of a single unit.

Table 10.4 Producer Risk as the Number of Units Increases.

Number of units	SD(W) in dollars
1	3,024,000
2	5,685,940
3	8,340,025
4	10,991,946
5	13,642,965
6	16,293,522
7	18,943,812
8	21,593,932
9	24,243,939
10	26,893,865

Table 10.5 Risk of Hedged Portfolio as Number of Units Increases

Number of units	$\sqrt{1-\text{Corr}(W, B)^2}$	SD($W + e^*(f - B)$) in dollars
1	.482	1,457,462
2	.363	2,061,163
3	.303	2,524,398
4	.266	2,914,924
5	.239	3,258,984
6	.219	3,570,039
7	.203	3,856,082
8	.191	4,122,325
9	.180	4,372,386
10	.171	4,608,900

- The earnings of the portfolio do increase linearly with the number of units. Since the risk increases less than linearly, the ratio of the risk to earnings decreases as the number of units increases.

We next investigate the risk of the hedged portfolio as the number of units increases. Table 10.5 provides the results.

We complete this example with some further remarks.

- As with the risk of the unhedged position, the increase in risk is nonlinear. Indeed, the risk of 10 units is substantially less than 10 times the risk of a single unit. This effect is more pronounced in the hedged portfolio than the unhedged portfolio.
- Also, similar to the unhedged portfolio, the risk in adding a single unit decreases as the number of units increases. The effect is far more pronounced in the hedged portfolio than in the unhedged portfolio. Indeed, the incremental risk of the tenth unit is only 236,514 dollars. This is 16 percent of the risk of a hedged single unit and only 8 percent the risk of an unhedged single unit.

10.5.6 Fuel and Power Price Uncertainty

We present one more example before summarizing results. The example addresses an additional complexity; both the gas price and ISO index are considered to be random variables. This example is included for completeness. A bit more sophistication is necessary to determine the optimal solution because optimization must be performed in two dimensions as opposed to one.

EXAMPLE 9. ***Uncertain Gas and Power Prices with a Single Unit***

The example follows the five-step method of determining the minimal risk portfolio for a single unit.

Step 1. Develop a set of simplifying assumptions.

The simple case that is investigated applies the following assumptions.

- There is only one unit, and the unit is a CCGT.
- The operational parameter H is constant.
- We ignore several terms and variables: V, g, $(D - I)M$, and the value S.
- We assume that the unit does not sell ancillary services into the market, $A = 0$.
- We assume that the value of capacity, K, is a constant.
- $P = B$.
- There are no forced outages.

Neglecting constant terms and including only those terms that are subject to risk, the future value of the portfolio becomes the following:

$$W = (P - HG)M$$

It is advantageous to rewrite the expression for W as follows.

$$W = aP + bG + E(W) + Z$$

where

- a and b are constants
- $E(W)$ is the expected value of W and is also a constant
- Z is a random variable that is uncorrelated from both P and G. Note that $E(Z) = 0$.

In the case that both H and M are constants, $a = M$, $b = -HM$ and $Z = 0$. For the more general case we apply the decomposition formula, Formula 7 of Section C. In this formula, a and b are regression coefficients given by the following expression.

$$a = \frac{\mathrm{Var}(G)\mathrm{Cov}(W,P) - \mathrm{Cov}(P,G)\mathrm{Cov}(W,G)}{\mathrm{Var}(P)\mathrm{Var}(G) - (\mathrm{Cov}(P,G))^2}$$

$$a = \frac{\mathrm{Var}(P)\mathrm{Cov}(W,G) - \mathrm{Cov}(P,G)\mathrm{Cov}(W,P)}{\mathrm{Var}(P)\mathrm{Var}(G) - (\mathrm{Cov}(P,G))^2}$$

This expression is a multivariable version of previous cases. One could create a scatter plot of the earnings, W, as a relation between power and gas. This would be a plot in three dimensions. The expression, $Q = aP + bG + E(W)$ is an equation for the plane that best approximates the scatter plot. Z is the difference between the scatter plot and the plane.

$$Z = W - Q$$

Step 2. Determine the set of market instruments that are incorporated in the portfolio.

The analysis requires consideration of fuel costs and interaction with fuel markets. Gas market instruments are the similar to those described for the power markets. We assume a daily gas index exists for a single gas delivery point into the ISO and that the price of this index is set the day before delivery. The market instruments are the following.

1. Fixed for floating contract using the daily gas index as the floating price. The fixed price of this contract is the market price for gas.
2. Monthly call options
3. Daily call options
4. Monthly put options
5. Daily put options

Assume that the producer wishes to financially mitigate risk by selling fixed for floating power contracts and purchasing fixed for floating gas contracts. Introducing the fixed for floating instruments for both power and gas yields the following.

$$W + e(f - B) + d(G - q) = aP + bG + E(W) + Z + e(f - B) + d(G - q)$$

With the assumption $B = P$, the expression for the portfolio becomes the following.

$$W + e(f - P) + d(G - q) = (a - e)P + (b + d)G + ef - dq + E(W) + Z$$

Step 3. Write down the expression for the variance of the portfolio with market instruments included and develop further simplifying assumptions.

Using Formulas 1, 2, and 3 from Section 10.3, the variance becomes the following:

$$\begin{aligned}\mathrm{Var}(W + e(f - B) + d(G - q)) &= (a - e)^2\mathrm{Var}(P) + (b + d)^2\mathrm{Var}(G) \\ &\quad + (a - e)(b + d)\mathrm{Cov}(P, G) + \mathrm{Var}(Z)\end{aligned}$$

Note that there are no other terms because Z is uncorrelated from both P and G.

Step 4. Solve for the quantity of the market instruments that minimizes the risk.

The variance is minimized by setting $e^* = a$ and $d^* = -b$.
With this solution,

Step 5. Determine the value of the risk associated with the optimal decision variables.

Using the optimal hedge, the standard deviation of W is equal to the standard deviation of Z.

$$\mathrm{SD}(W) = \mathrm{SD}(Z)$$

In the case that H and M are constants, $e^* = M$ and $d^* = HM$, $Z = 0$ and the portfolio is completely hedged, SD(W) = 0. In general the expression for the portfolio risk, SD(Z), is quite complicated and we leave the complete expression as an exercise for the reader. We do remark that the result approaches zero as the variance in both H and M approach zero. The result once again demonstrates the importance of minimizing uncertainty in H and M by proper maintenance and operating procedures that minimize operational uncertainty.

We conclude this section by discussing commonalities and differences between retailer risk and producer risk. One point in common is that market instruments do not address many of the risks of retailer and producer portfolios. Additionally, the instruments address ISO index risk incompletely. Table 10.6 summarizes the shortcomings of market instruments.

Another feature that is common to both retailers and producers is that operational performance has a large impact on risk. Poor load forecasting increases retailer risk, while excellent load forecasting decreases retailer risk. For the producer, fewer forced outages and a more stable output through diversification both decrease risk.

A significant difference between retailers and producers is that producers are able to manage their volumetric risk through unit diversification and sound operating and maintenance procedures. However, the retailer is unable to address volumetric risk due to customer departure. The inability to manage the risk of customer departure is a reason that retailers are less disposed toward participating in midterm electricity markets.

Table 10.6 Shortcomings of Market Instruments

Risk	Retailer	Producer
Volumetric risk	Market instruments do not hedge the risk associated with uncertain customer load.	Market instruments do not hedge the risk associated with uncertain unit output.
Price risk	Profile of market instruments does not match customer load.	Profile of market instruments does not match unit dispatch
Capacity volumetric risk	Uncertainty in capacity requirement	Uncertainty in capacity availability. May face unavailability penalty.
Other risks	No market instruments to address intraday market costs, imbalance penalties	No market instruments to address intraday market value, imbalance penalties, ancillary service value, uncertainty in operating expenses

The liquidity of power markets in the US is far below industry forecasts of the mid 1990s. A primary reason that the markets have not developed is the asymmetry of benefits to producers and retailers. Clear benefits of market participation only occur for one side, and it is not feasible to maintain one-sided markets.

The structure of the market is not the one that was envisioned by market designers. There has not been large-scale segregation of retailer and producer activities. Producers are unable to hedge in the midterm markets because there are few counterparties to sell to. One viable strategy is to develop their own retailer capabilities; customers act as counterparties to whom they can sell their output. Of course, they run the risk of customer uncertainty, but an uncertain customer base is preferable to a completely unhedged position. This strategy holds other advantages, as investigated in the next section.

10.6 RISK MANAGEMENT IN MIDTERM MARKETS: INTEGRATED ELECTRICITY SUPPLIERS

In this section, the risk position of an integrated electricity supplier that is both a producer and retailer is explored. It is seen that some risks between producers and retailers are complementary, with the consequence that the overall risk to the integrated company is lower than the risk of each individual company. The approach is similar to that of the integrated utility. The integrated electric supplier balances demand with its supply portfolio.

The future value of an integrated company's position is the sum of its retail and producer positions.

$$R + W = L(p - P(L) - n) - D - I - K + \sum_{i=1}^{n} W_i$$

$$W_i = M_i(P(M_i) - H_i(G + g_i) - V_i) + D_i - I_i + A_i + K_i - (S_i + z_i)$$

For completeness the symbols are once again defined:

- R is the retail component of the integrated portfolio in dollars.
- W is the producer component of the integrated portfolio in dollars.
- L is the retail load in MWH.
- p is the averaged end customer price in \$/MWH.
- $P(L)$ is the load-averaged ISO day-ahead index in \$/MWH.
- D is the cost of day-ahead adjustments to the ISO dispatch schedule.
- I is the ISO imbalance penalty.
- n is the ISO network charge.
- K is the cost to meet the capacity requirement.
- W_i is the future value associated with unit i.
- M_i is the total production of unit i in MWH.

- $P(M_i)$ is the unit i production-averaged ISO index in \$/MWH.
- H_i is the production-averaged heat rate for unit i.
- G is the day-ahead gas index.
- g_i represents transportation charge to bring fuel to unit i.
- V_i is the variable operations and maintenance charges in \$/MWH.
- D_i is the value of day-ahead adjustments to the ISO dispatch schedule of unit i in dollars.
- I_i is the ISO penalty for imbalances associated with unit i in dollars.
- A_i is the value of ancillary services that unit i receives in dollars.
- K_i is the capacity benefit of unit i earned by selling capacity tags in dollars. Note that there is a penalty associated with unavailability due to forced outage.
- S_i is the uncertain charges in operating and maintaining the generation units and includes the cost of unscheduled outages.
- z_i is the fixed charge of maintaining generation unit i and includes scheduled maintenance charges as well as debt servicing.

One effective way for the integrated company to mitigate risk is to develop a customer base that is optimally fit to its production base. The optimal fit requires that load volume and profile match the production volume and profile. This is formulaically expressed as follows:

$$L = \sum_{i=1}^{n} M_i = M_{\text{total}}$$

$$P = \sum_{i=1}^{n} \frac{P_i M_i}{M_{\text{total}}}$$

The first equation is straightforward; it equates the overall load with production volume. The second equation equates the price shape of the load with the price shape of production. Together these two equations define the conditions for the optimal fit. Suppose that it is possible to attain the optimal fit. Then the integrated portfolio becomes the following:

$$R + W = L(p - n) - D - I - K + \sum_{i=1}^{n} W_i$$

$$W_i = M_i(-H_i(G + g_i) - V_i) + D_i - I_i + A_i + K_i - (S_i + z_i)c_i$$

The price risk has been mitigated. Other risks are also offset as follows:

- Intraday price risk $\sum_{i=1}^{n} D_i - D$ is close to 0. Because load and supply are near equilibrium, intraday scheduling activities offset one another.

- Imbalance penalties and ancillary services: $\sum_{i=1}^{n} A(i) - I(i) - I$ is close to 0. The volume of ancillary services sold into the market must balance the imbalance penalties of retail and producer positions. The volumetric balance results in near financial balance. An integrated company with load and production in near equilibrium finds that its ancillary service capability and imbalance penalties are also in near equilibrium.
- Capacity benefits: $\sum_{i=1}^{n} K(i) - K$ is close to 0 for the same reason as above.

Assuming that these risks are completely offset, the integrated position becomes the following:

$$\begin{aligned} R + W &= L(p - n) + \sum_{i=1}^{n} M_i(-H_i(G + g_i) - V_i) - (S_i + z_i)c_i \\ &= L(p - n - H_{\text{ave}}(G + g_{\text{ave}}) - V_{\text{ave}}) - (S_{\text{total}} + z_{\text{total}})c_{\text{total}} \end{aligned}$$

where the subscript ave indicates average costs over the referenced component and the subscript total indicates total costs over the referenced component.

In the above equation the income stream is L_p and the remaining terms are the costs to serve the load, L. In the end, the customer must pay all the costs of service; L_p is larger than the remaining terms. Excluding transmission, this portfolio is precisely the portfolio of a traditional utility. Midterm planning for a traditional utility is the minimization of the cost terms in the above equation subject to meeting load and scheduling of maintenance requirements.

If one can pull into equilibrium before delivery, the main risk that must be mitigated through the markets is the risk in the fuel cost, $LH_{\text{ave}}G$. This risk contains heat rate risk, load risk, and gas price risk. With methods similar to those in the previous sections, it is possible to construct the optimal gas hedge.

Summary

There are several advantages that an integrated retailer has over producers and retailers. We list these advantages and close with other key points.

- The markets do not provide instruments that completely address risks that retailers and producers confront. The integrated supplier can mitigate several of these risks by offsetting positions.
- Illiquidity in midterm markets leaves producers with difficulties in hedging their portfolios. The integrated supplier addresses this weakness by developing a customer base that is a natural fit with its portfolio of power plants.
- By proper development and management of their portfolios along with sound operations, integrated suppliers are less likely to face customer departure than retailers. This is because integrated suppliers are able to align their production

portfolios with that of the market so that they are always able to bring their production to market at competitive prices.

- With these advantages, it is not surprising that incumbent utilities have maintained their dominant positions in markets that have opened up to free choice.
- While there are some examples of successful retailers, this is the exception rather than the rule. The market share of the successful retailers is extremely small in comparison to that of integrated suppliers.
- Integrated suppliers address volumetric risk just as retailers and producers. Sound load forecasting capabilities allow retailers to minimize volumetric risk. Sound operating procedures and maintenance practices along with large-scale ownership of units allows the integrated supplier to manage volumetric production risk.
- Given the importance of ownership of large-scale systems in managing volumetric risk, it is not surprising that most integrated suppliers as well as IPPs are very large corporations with excellent access to capital markets.

Chapter 11

The California Experience

California was the first state in the US to establish a nontraditional market environment allowing end customers to choose their retail service provider. This was done with much fanfare back in 1998. Proponents of the new market process claimed that competition would discipline both producers and retailers and consumers would benefit. The actual outcome was quite different and contained many ironies, including the following.

1. Utilities sold off 60% of their generation to IPPs with the expectation that many customers would leave. Later, utilities found themselves purchasing electricity generated from the very same units they sold at prices that exceeded 30 times the production cost so that they could service the customers that didn't leave.
2. A regulated rate regime was established for incumbent utilities to promote customer switching and assist utilities with cost recovery in a dwindling customer base. The regulated rate put one utility, PG&E, into bankruptcy and another utility, SCE, into near bankruptcy as purchases in the day-ahead markets exceeded the regulated rate 10-fold. The third major utility, SDG&E, remained healthy because it was allowed to abandon the regulated rate and increase prices dramatically.
3. Few customers chose to leave their incumbent utility. Almost all who did leave returned to the utility under a provider of last resort provision. The utility became the provider of last resort as prices from alternative retailers soared well above the regulated rate.
4. Producer behavior was not competitive, but anticompetitive. Producers gladly reduced their volumes to maintain high prices.

As is well known, California suspended its experiment. This chapter describes the market design and participant activities that led to the outcome in California. In Section 11.1, basic market principles are presented. Section 11.2 provides a technical description of California's market structure. The technical description provides a basis for understanding the trading tactics of Enron and other power marketers. In Section 11.3, we briefly discuss Fatboy, Get Shorty, and other trading tactics that

Electric Power Planning for Regulated and Deregulated Markets. By Arthur Mazer

exploited flaws in California's market design. Section 11.4 provides a contrast between the California design and PJM's market design as described in Chapter 7.

11.1 MARKET FUNDAMENTALS

The ideological bent of the market designers was that of free market economists. The features of the market indicate that their main focus was to ensure that a price signal would be available to all parties so that the parties could respond to the price signal. The common belief was that a price signal would indicate the market value of electricity that both consumers and producers could respond to. Consumers would reduce consumption if the price became too high, while competitive suppliers would respond to favorable price signals by providing more power. There are arguments to be made for and against this viewpoint. The arguments are a bit moot because this guiding principle was never fully implemented. The market design was set in a political environment with competing stakeholder interests. Decision making was managed through consensus and compromise. In this process the central concept of price response was compromised away. Although the price signal was publicly available, it was rendered meaningless by a regulated price mechanism that is explained further below.

The California market designers segregated the power industry into two components, network issues and supply. The desire was to move in a direction of regulating all networking issues while allowing free markets to govern supply. The vision was a field of IPPs and retailers competing in free markets with fair access to the grid overseen by an ISO.

To this end, the California Public Utilities Commission (CPUC) created the California Independent System Operator (CAISO) as a public agent and tasked CAISO with overseeing all physical issues regarding the grid. The CPUC established another public agency, California Power Exchange (CALPX), as a financial clearinghouse for transactions between IPPs, retailers, and utilities. Through the CALPX and CAISO, short-term dispatch and pricing was set. CALPX also ran a forward market for mid-term transactions, but market participation was very low.

An interesting point to note is that the CPUC had oversight authority over the utilities, SCE, PG&E, and SDG&E, but not over any other market participants. This authority allowed the CPUC to require that utilities manage their short-term activities through CALPX and CAISO. Parties that did not fall under CPUC authority were not subject to CPUC regulation. For example, the Los Angeles Department of Power and Water, (LADPW) a local authority that provides power to the city of Los Angeles, chose not to join CAISO. LADWP continues to operate the grid throughout its control area. The city of Los Angeles actually benefited from the California energy crisis, as it sold surplus electricity from its own supply resources into the CALPX at very profitable prices without experiencing any power disruptions within its own service territory.

The market structure was a political compromise between many parties, including the CPUC, the utilities, energy marketers, industrial consumers, and various residential consumer constituencies. Below we present some of the features that emerged from the diverse political mix.

For the utilities, stranded cost was a big issue. Utilities commonly enter into long-term contractual obligations with CPUC approval. The CPUC also provides utilities with provisions to recover the costs of such obligations through the regulated rate mechanism. In a market with choice, these provisions would no longer apply; utilities would be saddled with long-term obligations but would not have a fixed customer base or regulated rate mechanism to recover these costs. Without a new provision to recover stranded costs, utilities would not agree to market restructuring.

The solution to the stranded cost issue was a regulatory mechanism that introduced regulated pricing into the new market design from the outset. The CPUC agreed to a regulated rate mechanism that initially fixed utility customer prices until the utility's stranded costs were recovered. After cost recovery, there was to be no regulated price. It was foreseen that the process might take 5 years.

Aside from providing accelerated cost recovery, there was another motive for the rate increase. It was believed that this would assist the entry of new market players wishing to become retailers. Entry of such players was desired in order to have competitive choice.

A frequently sited shortcoming of the California design was that the CPUC severely restricted utility participation in the CALPX forward market where midterm transactions were cleared. With limited utility participation, there was little activity in the midterm markets. The reason for this prohibition was the consensus that utilities were too dominant and they would be able to manipulate markets with their dominant positions. Whether or not greater utility participation in forward markets would have led to a different result is a subject of debate.

Another feature of the California market was the supplier of last resort clause. In regulated markets the utility has the obligation to serve every customer in its service territory. Everyone in the service territory is covered. In a deregulated market there is no such protection. If a retailer defaults and stops providing service, a customer might be without service. On the other side, a retailer might restrict its customer base to a targeted group and those outside the target may have difficulty obtaining service. To ensure guaranteed service across the population, utilities were obligated to take on any customer as the supplier of last resort.

The California experiment had a short life that can be described as follows.

1. March 31, 1998. The CALPX opens.
2. Utilities sell off a significant percentage of their generation, 60%, to IPPs while maintaining a significant percentage of their customer base and charging them the regulated rate of around \$.065/kWH (\$65/MWH).
3. IPPs exercise market power. They sell into the CALPX at over \$1000/MWH.
4. Most retailers default, and utilities pick up their customers as the supplier of last resort. Utilities were obligated to service their customer base without interruption regardless of market prices.

5. PG&E declares bankruptcy. Its cash resources have been depleted by buying power at the CALPX and selling it to customers at the regulated rate.
6. January 2001. The CALPX ceases operations, ending the experiment.

The critical element that caused the collapse was Point 2. Had the utilities not sold off a significant amount of generation and moved dangerously far from equilibrium, the IPPs would never have been able to exercise market power.

11.2 SHORT-TERM MARKET STRUCTURE: THE CALPX, CAISO, AND OTHER MARKET PARTICIPANTS

In this section we describe how the day-ahead through real-time California market functioned. The format follows Section 7.2, where a description of the short-term activities in the hypothetical market is presented. In that section and below, operational requirements are matched with the responsible market participant. However, differences in presentation result from a difference in market design. The hypothetical market centralizes dispatch and grid operations within a single agent, the ISO. California's market design segregated responsibilities, which resulted in fractured markets. Indeed, there were three distinct markets, one run by CALPX, and two run by the ISO. This section provides distinct subsections for all three of these markets.

The CALPX market was an energy-only market. It established a day-ahead and intraday market clearing price for energy. Although the initial vision was to execute transactions on an hourly basis during delivery date, this became infeasible. Instead, the market transacted at less frequent intervals throughout delivery day. CAISO operated an ancillary services market that was separate from the CALPX market. However, as with the CALPX market, transactions were executed the day ahead and at scheduled intervals during the delivery day. Additionally, CAISO operated a real-time market. These are the three markets that we describe.

In addition to its responsibility of clearing energy bids, CALPX also assumed the role of scheduling coordinator on behalf of the utility and merchant energy companies that participated in the CALPX market. A scheduling coordinator coordinated physical scheduling with CAISO for both generation and load. The scheduling coordinator had three responsibilities: submitting generation dispatch schedules, submitting load schedules, and scheduling transmission. CALPX submitted load and dispatch schedules only for the energy that was cleared through the CALPX market. Not all energy within CAISO's territory was cleared through CALPX; indeed, merchant energy companies such as Enron would submit their own schedules to CAISO for energy that they did not clear through CALPX. Accordingly, there were many scheduling coordinators. One CAISO requirement was that all scheduling coordinators submit balanced schedules: Scheduled generation had to match load requirements except for power scheduled to be exported outside of CAISO.

Before proceeding to describe the market mechanics, we reiterate that the design had a critical flaw: There was not a single agent setting the dispatch and operating the grid. In Chapter 3 the importance of centralizing these activities is stressed. The well-publicized tactics of power marketers such as Enron took full advantage of this flaw and are discussed in the next section.

11.2.1 CALPX Energy Market

This subsection provides a list of activities and division of responsibilities within the CALPX energy market.

1. Day-Ahead Load Forecasting and Demand Bidding

This aspect of day-ahead operations worked nearly the same as the market design of Chapter 7, with the exception that the market organizer was the CALPX, not the CAISO. Both retailers and the CAISO had a stake in the load forecast and accordingly performed their own forecasts. The retailers determined their requirements in a day-ahead forecast and informed the CALPX of their day-ahead procurement requirements. The CAISO also developed its own forecast so that it could arrange appropriate resources for operating the grid.

Retailers provided the CALPX with load schedules that are the same as those described in Chapter 7. Table 7.1 is reprinted below.

2. Day-Ahead Supply Offers into Energy Markets

Producers provided the CALPX with the price of their supply resources. The form of bidding varied greatly from that of Chapter 7, where offers were entered as operational parameters. In the CALPX market, suppliers provided an hourly schedule of prices and volumes similar to Table 11.1.

3. CALPX Unconstrained Day-Ahead Hourly Price and Supply Request

The market clearing price was set by the exact process that is illustrated in Section 4.3. There, a dispatch schedule is set by stacking up the units in merit order and selecting the units from least costly to most costly until the forecasted load is satisfied. The CALPX stacked up the supply bids in merit order and accepted the bids

Table 7.1 Retailer's Bid Table

Hour ending 2 P.M.	
Price in $/MWH	Demand in MWH
<80	3800
80 to <175	3700
175 to <250	3550
≥250	3350

Table 11.1 Supply Bid

Hour ending 2:00 P.M.	
Total MWH	Price in $/MWH
450	25.00
855	27.25
1005	32.00
1105	38.50
1200	45.00
1250	75.00

until the demand as requested from Step 1 was satisfied. The last supply bid chosen set the market clearing price.

Recall that the merit order selection in Chapter 4 did not take into account operational requirements of either the units or the grid. The resulting price was an unconstrained price in the entire CAISO control area as this was considered a single territory without transmission constraints.

After setting the market clearing price, the CALPX notified each producer of its quantities that cleared the CALPX market at the unconstrained price. This notification was a financial commitment between CALPX and the producers.

4. Provisional Setting of the Dispatch

Once producers received the quantities that CALPX would clear, they set a dispatch schedule for their units. The producers would convey the dispatch schedule of units that were located in the CAISO control area as well as import schedules from resources outside of CAISO's control area; imports were a significant volume of supply.

Note that this schedule was an unconstrained schedule that had not been subject to a power flow analysis. In general, it was not an operationally feasible schedule. However, this was a financially binding schedule between the producers and the CALPX.

5. Schedule Adjustment Bidding

Along with their dispatch schedule, suppliers would provide bids for deviating from the provisional schedule. Deviations were necessary to produce a final feasible schedule. Two sets of bids were submitted, inc and dec bids. Inc bids were the price for increasing output, and dec bids were the price for decreasing output. These bids were associated with specific units as shown in Table 11.2.

6. Caiso's Schedule Adjustment and Congestion Price

Typically, the transmission transfer capabilities between northern and southern California were unable to accommodate the CALPX preliminary schedule. To make

Table 11.2 Inc and Dec Bids

Inc bid for "Unit Name" for hour ending 4 P.M.	
Increment by X MW	Price
10	10
20	15
30	25
Dec bid for "Unit Name" for hour ending 4 P.M.	
Decrement by X MW	Price
20	5
30	10
40	20

adjustments, CALPX sent the preliminary schedule along with adjustment bids to CAISO. CAISO would then optimally adjust schedules to bring the schedule in line with grid capabilities. This was accomplished by selecting the cheapest incremental and decremental bids necessary to redistribute the schedule so as not to overburden north-south grid capacity as well as other congested pathways. This adjustment determined a congestion price for use of the transmission system. The congestion price was the sum of the most expensive increment and most expensive decrement of the adjusted schedule.

7. Final Day-Ahead Schedule and Congestion Fee Settlement

CAISO sent its adjusted schedule and congestion price to CALPX. CALPX informed the producers of their schedule adjustments as well as their allocation of the congestion fee. Note that schedules on the source side of the transmission constraint were decremented while schedules on the delivery side of the transmission pathways were incremented. Producers were responsible for paying the congestion fee for use of the grid. The scheduled use of the north-south transmission pathways was prorated among the producers who were producing on the source side of the flow at the cost of the congestion fee. A producer's prorated volume was the percentage of the adjusted schedule that a producer delivered into the source side of the transmission pathways times the scheduled volume over the transmission pathway. For example, suppose that 15,000 MWH of power is being produced in the northern sector for the hour ending at 11 A.M. Suppose that 5000 MWH is scheduled to flow from north to south and a certain producer is scheduled to deliver 3000 MWH in the northern sector. Then for the hour ending at 11 A.M., the prorated share over the transmission pathways for the producer is 1000 MWH. If the market clearing inc and dec bids summed to \$21, the transmission price that the producer paid was \$21,000.

The effect of the transmission fee was that the inc and dec market clearing prices were paid by the producers on the source side of the transmission pathways to the

Table 11.3 Transactions Between CALPX, CAISO, Retailers, and Producers

Transaction type	Cash flow	Calculation
Energy purchase	Retailer to CALPX	Market clearing of energy price times volume
Energy sale	CALPX to producer	Market clearing of energy price times volume
Decrement of schedule	CAISO to CALPX CALPX to producer	Market clearing price of decrement times volume
Increment of schedule	CAISO to CALPX CALPX to producer	Market clearing price of increment times volume
Congestion fee	Producer to CALPX CALPX to CAISO	Congestion price times prorated volume of transmission path

curtailed producers on the source side of the transmission pathways as well as to the incremented producers on the delivery side of the congested transmission pathways.

At this point the day-ahead financial commitments between the market participants and CALPX were set. Table 11.3 lists all transactions between CALPX and market participants. Note that all financial flows pass through the CALPX; however, CALPX receives no benefit, as all incoming payments match outgoing payments. Cash flows related to the transmission congestion fee passed through CAISO since CAISO was the grid operator. As seen in the table, net revenues to CAISO were zero since the congestion fee equals the sum of the increment and decrement fees.

The above steps were repeated several times during delivery day.

11.2.2 CAISO's Ancillary Services Market

1. Day-Ahead Setting of AS Requirements

In California, there was an AS market for four services: regulation, primary spinning reserves, primary nonspinning reserves, and secondary reserves. CAISO managed other necessary services such as providing supplementary reactive power and voltage support through a regulatory mechanism. CAISO determined the AS requirements in accordance with the standards of WECC, the regional reliability council, and CAISO's forecast of needs.

2. Day-Ahead Supply Offers into Reserve and AS Markets

Generation owners bid into the AS markets. Producers could only bid into the AS market if they operated plants that were certified to provide such services. The bids for all reserve services contain three elements: a capacity quantity in MW, a capacity price in dollars, and an energy price in dollars per MWH.

3. Day-Ahead Setting of AS Prices

Using CAISO's forecast of AS requirements and the producer's AS bids, CAISO had a supply and demand outlook for ancillary services. Prices for these services could then be set at the intersection of the supply and demand curves.

4. Informing Suppliers of Their Status

CAISO would inform generators if they were selected or not.

In addition to operating the day-ahead AS market, CAISO would operate bidding sessions for ancillary services during delivery date.

Remark

- A regulatory mechanism was necessary for the procurement of certain grid services because it is not possible to create competitive bidding conditions for all grid requirements. Providing reactive power for local voltage support is an example of a grid service that may not be serviced by the number of producers required to create competitive conditions. In fact, there are units that must operate in order for the grid to function. The owners of such units have monopolies and are subject to regulated pricing.

11.2.3 CAISO Real-Time Market

In addition to the ancillary services market, CAISO operated a real-time market through its real-time dispatch decisions. CAISO optimized ancillary services and schedule adjustment bids to balance supply and demand in real time. The marginal energy cost of these services determined the real-time price, which the ISO published after the fact.

Every generator received the real-time price for generation beyond its scheduled generation. Generators that delivered less than their scheduled energy commitments paid the real-time price. Similarly, every load retailer paid the real-time price for every unscheduled MWH of consumption and received the real-time price for scheduled load that did not materialize.

11.3 FATBOY, GET SHORTY, RICOCHET, AND DEATH STAR

The title of this section gives the names of trading tactics that Enron used during the California energy crisis. We present these tactics as examples of how the market design was exploited.

11.3.1 Fatboy

Fatboy was a tactic that Enron used to take advantage of anticipated differences between the CALPX market and CAISO's real-time market. Enron traders adopted this tactic when they felt that the real-time market would settle higher than the CALPX market. As scheduling coordinator, Enron was subject to submitting balanced schedules to CAISO. Enron would submit inflated load schedules and schedule a corresponding amount of generation. When the load did not materialize in real time, Enron received the real-time payment for the load that did not materialize. This quantity was identical to the surplus generation in excess of Enron's true load requirement. Whenever Enron correctly guessed that the real-time price would be higher than the CALPX price, it would earn more than it would have had it bid its surplus production capacity into the CALPX market.

11.3.2 Get Shorty

Get Shorty was an inversion of the Fatboy tactic. Enron traders adopted this tactic when they felt that the real-time market would settle lower than the CALPX market. In this case, Enron would sell nonexistent power into the CALPX market. Enron was later charged the real-time price for power that it did not provide. Whenever Enron correctly guessed that the real-time price would be lower than the CALPX price, it would earn the difference between these two market prices.

Another version of Get Shorty was for Enron to claim that it had procured ancillary service provisions on behalf of its retail customers that it in fact had not procured. In this manner Enron would avoid paying CAISO's charge for ancillary services.

11.3.3 Ricochet

Ricochet was a tactic used to circumvent CALPX price caps that applied to generation within the state of California. A price cap of $250/MWH was instituted by the State of California in response to persistently high price levels. The State of California had no jurisdiction over interstate commerce; accordingly, the price caps did not apply to purchases of power from outside the state. Enron would purchase power from CAISO at the price cap and sell the power to a subsidiary out of state. The subsidiary would then resell the power back to California at a price well above the price cap, at times above $1000/MWH.

11.3.4 Death Star

Death Star was a way to collect the congestion fee from CAISO without providing congestion relief. Enron would schedule transmission flow going against the congested direction. CAISO awarded Enron the congestion fee since the schedule would theoretically free up an equivalent amount of transmission along the congested direction. Enron would then schedule power flow back to its original point of origin along a transmission path that was not controlled by CAISO, negating any congestion benefit. The cost for scheduling transmission flows along the pathway not controlled by CAISO was lower than the congestion fee. Enron profited by arbitraging this difference.

The political fallout from the crisis created a finger pointing environment in which these tactics were natural targets. On discovering documents that alluded to the above tactics, the Governor of California, Governor Davis, spoke of a smoking gun that pointed to Enron and other energy companies engaging in similar activities as the cause of the crisis.

A reexamination of the crisis indicates that the tactics did not cause the energy crisis. The resulting profits from these tactics were not even close to the energy costs resulting from the crisis that Californians paid and are projected to pay for several years into the future. These tactics were symptoms of a dysfunctional system. The system was dysfunctional on two levels. First, the utilities were far from equilibrium, placing California consumers at the mercy of the energy merchants. Energy merchants took advantage of their position to attain handsome profits within the legal framework of the market that was established. Second, the market design did a poor job of accounting for operational requirements of delivering power. The above tactics fully exploited the operational cracks.

11.4 MARKET CONTRAST: PJM AND CALIFORNIA

In this section we present a contrast between the California market and that of PJM. Central to the contrast is that PJM's market is designed around operational imperatives whereas the California market was designed to promote competitive bidding. The section addresses two fundamental differences between PJM and California's market design that resulted from the two philosophies. The first and most critical fundamental difference is that California market designers segregated the responsibilities for clearing the energy market and operating the grid, whereas PJM centralized these responsibilities within a single agent. The second difference is the design of the bidding process.

In California, CALPX cleared the market, providing the market price around which the dispatch was set. The market price did not take into account grid operations and resulted in infeasible preliminary schedules. Tactics described in the previous section took advantage of this faulty scheduling. Indeed, energy merchants were able to set phantom schedules that caused transmission congestion and to receive payments for mitigating phantom congestion that they created.

Aside from the CALPX there were many other scheduling coordinators, each of whom independently set a balanced schedule with CAISO. This process provides another example of an inefficiency caused by the segregation of scheduling duties and grid operations. Requiring each scheduling coordinator to individually provide balanced schedules did not allow for optimization of resources among the different scheduling coordinators. Each scheduling coordinator optimized among its resources to balance the system, and a global optimization was not achieved.

In California, the CALPX ran the day-ahead market while CAISO ran the ancillary service market. Since the two markets were not cooptimized by a single agent, there were pricing inconsistencies such as those caused by congestion. Market participants took advantage of the inconsistencies by using tactics such as those presented in the previous section.

By contrast, PJM tasks a single agent, the ISO, with the energy and ancillary service market clearing function, day-ahead dispatch scheduling, and real-time operations. This allows for optimal and feasible day-ahead dispatch scheduling. The centralization of market clearing activities and grid operations also provides for pricing consistency between the day ahead, ancillary service markets, and the real-time price. The pricing consistency does not allow for opportunities to game the market.

The hypothetical market introduced in Chapter 7 creates strong disincentives for providing inaccurate production and demand information through the bidding process. Stiff penalty fees apply for actual flows that deviate from the schedule that results from the bidding process. The penalties ensure that the ISO is well informed to be able to dispatch the system economically and maintain reliability. This contrasts with CAISO. In the California market, market participants frequently set schedules that they had no intention of maintaining. Fatboy, Get Shorty, Death Star, and other tactics that created false congestion were all based on passing incorrect information to the CAISO by setting phantom schedules. Aside from the additional congestion costs that this caused, this in all likelihood created a level of operational uncertainty that affected CAISO's management of the grid.

PJM does not rely on imbalance penalties to ensure that market participants bid accurate forecasts. The results have been favorable. The market has run smoothly without the bidding distortions that occurred in California.

Another difference between PJM and the California design is the relation between the pricing and bidding processes. PJM sets the market clearing price, using locational marginal pricing. This allows for the separation of pricing points in line with the transmission constraints of the grid. The California design first set an unconstrained price across the entire system. The resulting dispatch was not feasible and contributed to congestion problems.

While the California approach was inferior from an operational approach, designers believed that it would provide a more competitive bidding process. Their line of reasoning was as follows. Because actual grid constraints limit a producer's ability to supply power, imposing the grid constraints limits competition and leads to a higher market clearing price. Conversely, by setting an unconstrained bidding pool, competition is enhanced because all producers compete in the unconstrained market.

While competitive bidding is desirable, it cannot be achieved by pretending that transmission constraints don't exist. California's system did not result in more competitive bidding. The grid constraints were well known by all market participants. Some took advantage of the constraints to earn higher revenues by employing tactics that are described in the previous section.

The bidding process also differed in that generators in PJM bid in their operating parameters while generators in the California market bid in a price by hour and volume. The price bidding in the California markets provided a simple method for the CALPX to match load and production bidding. However, it was problematic for producers because production quantities and costs could only be estimated.

Producers most likely made conservative estimates to ensure full cost recovery. As an example, consider the allocation of the start-up cost of a CT into the price bid. Assume that the start-up cost is $6000 per start and that the capacity of the CT is 100 MW. Bidding into the CALPX was submitted on a $/MWH basis. Without knowing how many hours the CT would be scheduled, the producer could not properly allocate the start-up cost into its hourly bids. The producer had to make a guess at what the schedule would be. Suppose the producer guessed that the unit would be scheduled to provide maximum output for 10 hours; then the start-up cost could be allocated as a $6/MWH cost adder into the bid. However, if after scheduling, the unit was called upon for less than 10 hours, the producer would not recover the start-up cost. A producer would most likely make a much more conservative guess concerning the number of scheduled hours to ensure that the start-up cost would be recovered.

CALPX did not consider the operational characteristics of the units and set the initial schedule on market price alone. At times, CALPX's schedule did not match the operational characteristics of the units. For example, in accordance with the market clearing price a unit may have been scheduled for a particular hour. The following hour the market clearing price may have slipped below the unit's bid so the unit was not scheduled. The next hour the market clearing price may have once more increased beyond the bid and the unit was scheduled. A unit would not be able to meet such a schedule because of constraints on ramp-up and ramp-down rates, minimum downtimes, and minimum uptimes. This type of schedule did occur and was outside of any units' operating capability.

Within PJM there is no incompatibility between the dispatch schedule and units. Operational parameters are submitted alongside the bid and are accounted for by the ISO as it sets the schedule.

11.4.1 Implementation of Markets

Aside from the differences in market structure, PJM's implementation of its market was very different from California's. California encouraged its utilities to quickly sell off power plants by providing them with a regulated rate over a short time window, 5 years. The regulated rate was supposed to assist with stranded cost recovery. Utilities responded and indeed quickly sold off 60 percent of their generation assets. The result was that utilities were dangerously out of equilibrium and exposed to market manipulation.

On the other hand, PJM set a more deliberate cost recovery schedule to address stranded cost recovery. The result is that the incumbent utilities within PJM have not sold off the levels of capacity that California utilities had. The utilities within PJM maintain a healthy balance between generation capacity and customer load. This is critical because PJM did not experience the shifting of customers away from utilities toward alternative retail service providers as market designers anticipated. Instead, utilities serve as integrated energy companies and are dominant within their traditional service territories.

Bibliography

BAZARAA, M. S., JARVIS, J. J., and SHERALI, H. D., *Linear Programming and Network Flows*, John Wiley and Sons, Hoboken, New Jersey, 2005

BERGEN, A. R., *Power Systems Analysis*, Prentice-Hall, Englewood Cliffs, New Jersey, 1986

BOYLE, G. (editor), *Renewable Energy*, 2nd edition, Oxford University Press, Oxford, 2004

BREALY, R. A., and MYERS, S. C., *Principles of Corporate Finance*, McGraw-Hill, New York, 1996

CASAZZA, J., DELEA, F., *Understanding Electric Power Systems: An Overview of Technology and the Marketplace*, John Wiley and Sons, Hoboken, New Jersey, 2003

CREW, M. A., and KLEINDORFER, P. R., *The Economics of Public Utility Regulation*, MIT Press, Cambridge, Massachusetts, 1986

CROSNO, D. C., *Fundamentals of Electromechanical Conversion, Harcourt*, Brace and World, New York, 1968

CUSATIS, P., *Hedging Instruments and Risk Management*, McGraw-Hill, New York, 2005

DASH, J. W., *Quantitative Finance and Risk Management: a Physicist's Approach*, World Scientific, River Edge, New Jersey, 2004

DUFFIE, D., *Dynamic Asset Pricing Theory*, Princeton University Press, Princeton, New Jersey, 2001

DUNAR, A. J., *Building Hoover Dam: an Oral History of the Great Depression*, Twayne Publishers, New York, 1993

EECKHOUDT, L., GOLLIER, C., and SCHLESINGER, H., *Economic and Financial Decisions Under Risk*, Princeton University Press, Princeton, New Jersey, 2005

ELGERD, O. I., *Electrical Systems Theory: An Introduction*, 2nd edition, McGraw-Hill, New York, 1982

EYDELAND, A. and WOLYNIEC, K., *Energy and Power Risk Management*, John Wiley and Sons, Hoboken, New Jersey, 2003

FEYNMAN, R. P., LEIGHTON, R. B., and SANDS, M., *The Feynman Lectures on Physics*, Addison-Wesley, Reading, Massachusetts, 1965

FREUND, J. E., *Modern Elementary Statistics*, 10th edition, Prentice-Hall, Upper Saddle River, New Jersey, 2001

GOMEZ, T., *Electricity Economics: Regulation and Deregulation*, John Wiley and Sons, Hoboken, New Jersey, 2003

HEIER, S., *Grid Integration of Wind Energy Conversion Systems*, John Wiley and Sons, Hoboken, New Jersey, 2006

HILLIER, F. S. and LIEBERMAN, G. J., *Introduction to Operations Research*, 8th edition, McGraw-Hill, New York, 2005

HULL, J. C., *Options, Futures and Other Derivatives*, 6th edition, Prentice-Hall, Upper Saddle River, New Jersey, 2006

Electric Power Planning for Regulated and Deregulated Markets. By Arthur Mazer

HUNT, S., *Making Competition Work in Electricity*, John Wiley and Sons, Hoboken, New Jersey, 2002

KEHLHOFER, R., *Combined Cycle Gas and Steam Power Plants*, PennWell, Tulsa, Oklahoma, 1999

KIRSCHEN, D. S. and STRBAC, G., *Fundamentals of Power System Economics*, John Wiley and Sons, Chichester, England, 2004

LAFONT, J. J., *A Theory of Incentives in Procurement and Regulation*, MIT Press, Cambridge, Massachusetts, 1993

LILLINGTON, J. N., *The Future of Nuclear Power*, Elsevier, Amsterdam, 2004

MCLEISH, D. L., *Monte Carlo Simulation and Finance*, John Wiley and Sons, Hoboken, New Jersey, 2005

MCNAMARA, W., *The California Energy Crisis: Lessons for a Deregulating Industry*, PennWell, Tulsa, Oklahoma, 2002

MEIER, A., *Electric Power Systems: a Conceptual Introduction*, Wiley-Interscience, Hoboken, New Jersey, 2006

MILLER, B. G., *Coal Energy Systems*, Elsevier Academic Press, San Diego, California, 2005

NEWBERY, D. M. G., *Privatization, Restructuring and Regulation of Network Utilities*, MIT Press, Cambridge, Massachusetts, 1999

RATIU, T. S. and MARSDEN J. E., *Introduction to Mechanics and Symmetry*, 2nd edition, Springer, New York, 1999

RAU, N. S., *Optimization Principles: Practical Applications to the Operation and Markets of the Electric Power Industry*, Wiley-Interscience, Hoboken, New Jersey, 2003

ROSS, S., *A First Course in Probability*, 6th edition, Prentice-Hall, Upper Saddle River, New Jersey, 2002

SACCOMANNO, F., *Electric Power Systems: Analysis and Control*, Wiley-Interscience, Hoboken, New Jersey, 2003

SHAHIDEHPOUR, M., YAMIN, H., and LI, Z., *Market Operations in Electric Power Systems*, John Wiley and Sons, Hoboken, New Jersey, 2002

SHAHIDEHPOUR, M. and ALOMOUSH, M., *Restructured Electrical Power Systems: Operation, Trading and Volatility*, Marcel Dekker, New York, 2001

STAGLIANO, V. A., *A Policy of Discontent*, PennWell, Tulsa, Oklahoma, 2001

STEVENSON, W. D. Jr., *Elements of Power Systems Analysis*, 4th edition, McGraw-Hill, New York, 1982

STOFT, S., *Power System Economics: Designing Markets for Electricity*, Wiley-Interscience, Hoboken, New Jersey, 2002

TAPIERO, C. S., *Risk and Financial Management: Mathematical and Computational Methods*, John Wiley and Sons, Hoboken, New Jersey, 2004

VARIAN, H. R., *Intermediate Microeconomics: A Modern Approach*, 5th edition, W. W. Norton, New York, 1999

VINCENT, T. L. and GRANTHAM, W. J., *Nonlinear and Optimal Control Systems*, Wiley-Interscience, Hoboken, New Jersey, 1999

WARKENTIN-GLEN, D., *Electric Power Industry in Nontechnical Language*, 2nd edition, PennWell, Tulsa, Oklahoma, 2006

WEBER, C., *Uncertainty in the Electric Power Industry: Methods and Models for Decision Support*, Springer, New York, 2005

WEHR, K., *America's Fight Over Water: the Environmental and Political Effects of*

Large Scale Water Systems, Routledge, New York, 2004

WOLLENBERG, B. F. and WOOD, B. F., *Power Generation and Operational Control*, 2nd edition, John Wiley and Sons, New York, 1996

WOODRUFF, E. B., LAMMERS, H. B., and LAMMERS, T. F., Steam Plant Operation, McGraw-Hill, New York, 1998

OTHER SOURCES

The US-Canada Power System Outage Task Force, *Final Report on the August 14th, 2003 Blackout in the United States and Canada*, 2004

TVA website: www.tva.com

PJM website: www.pjm.com

PBS website: www.pbs.org

Index

Page references followed by t indicate material in tables.